Cultural Approaches to the History of Medicine

*Other books by the editors*

## Willem de Blécourt

HET AMAZONENLEGER. *Irreguliere genezeressen in Nederland 1850–1930* (An Army of Amazones: Irregular Women Healers in the Netherlands, 1850–1930)

WITCHCRAFT AND MAGIC IN EUROPE, vol. 6: *The Twentieth Century* (*with Ronald Hutton and Jean La Fontaine*)

BEYOND THE WITCH TRIALS (*co-editor with Owen Davies*)

WITCHCRAFT CONTINUED (*co-editor with Owen Davies*)

## Cornelie Usborne

THE POLITICS OF THE BODY IN WEIMAR GERMANY: Women's Reproductive Rights and Duties

GENDER & CRIME IN MODERN EUROPE (*co-editor with Margaret Arnot*)

# Cultural Approaches to the History of Medicine

## Mediating Medicine in Early Modern and Modern Europe

Edited by

Willem de Blécourt and
Cornelie Usborne

Foreword © the Estate of Roy Porter 2004
Editorial matter and selection © Cornelie Usborne and Willem de Blécourt 2004
Chapters 1–13 © Palgrave Macmillan Limited 2004

First published 2004 by
PALGRAVE MACMILLAN
Houndmills, Basingstoke, Hampshire RG21 6XS and
175 Fifth Avenue, New York, N. Y. 10010
Companies and representatives throughout the world

PALGRAVE MACMILLAN is the global academic imprint of the Palgrave Macmillan division of St. Martin's Press, LLC and of Palgrave Macmillan Ltd. Macmillan® is a registered trademark in the United States, United Kingdom and other countries. Palgrave is a registered trademark in the European Union and other countries.

ISBN 1–4039–1569–5

This book is printed on paper suitable for recycling and made from fully managed and sustained forest sources.

A catalogue record for this book is available from the British Library.

Library of Congress Cataloging-in-Publication Data
Cultural approaches to the history of medicine: mediating medicine in early modern and modern Europe/edited by Willem de Blécourt & Cornelie Usborne.
    p. cm
Includes bibliographical references and index.
ISBN 1–4039–1569–5
    1. Social medicine–Europe–History. 2. Medicine–Europe–History.
3. Medical care–Europe–History. I. Blécourt, Willem de. II. Usborne, Cornelie, 1942–

RA418.3.E85C85 2003
306.4'61'094–dc21                                             2003046923

10    9    8    7    6    5    4    3    2    1
13    12    11    10    09    08    07    06    05    04

Printed and bound in Great Britain by
Antony Rowe Ltd, Chippenham and Eastbourne

*Dedicated to the memory of*
*Roy Porter (1946–2002)*
*who inspired a whole generation of historians*

# Contents

# List of Illustrations

# More than a Foreword

*Roy Porter*

Heroic tales of great doctors making brilliant breakthroughs have long since had their day, at least in the history of medicine as an academic discipline. But what replaced them from the 1980s is equally looking no less dated and distorted. No longer does it seem adequate to speak simply and brashly of 'medical imperialism', the imposition of medical values; or of doctors as the new priesthood in an unholy alliance with the state; or of the naturalization and biologization of politico-moral prejudice in the name of biomedicine. The era in which dropping the name of Illich, Althusser or Foucault could clinch an argument, or in which bold but simplistic arguments about the 'medical control' of women, sexual minorities or ethnic groups, is over. Indeed, 'new presbyter' once again proved 'old priest writ large', and later orthodoxies have become targets for new historians of medicine, who feel driven to transcend the recently fashionable dogmas of social constructivism. By way of exploration of detailed case studies drawing upon archival materials, and in the light of a new multicultural pluralism, sensitive to but not scripted by particular theoretical constructs, today's historians are seeking to construct richer, less opinionated accounts of medicine and its roles in the past.

*Mediating Medicine* is a welcome pioneering contribution to this trend. What is commendable about it is that it is not an account of medicine from the doctor's point of view, from science's point or view, from the patient's point of view, or, for that matter, from society's point of view. Rather it is an attempt to reconstruct the subtle webs of people and policies which have gone to make up the weave of medicine in the past. As such, it shows a deep commitment not to scoring polemical points or to unveiling 'hidden forces' at work in the past, but to recovering and establishing the meanings of such developments.

One of the most valuable aspects of the essays in this collection lies in their reconstruction of the complexity of the worlds of medicine in the past, in particular before the nineteenth-century trend towards professionalization and rigid legislative control of the medical profession. For example, in 'Seventeenth-century English Almanacs: Transmitters of Advice for Sick Animals', a pioneering study of almanacs in seventeenth-century England, Louise Hill Curth draws upon the vast but neglected evidence of popular almanacs, produced on an annual basis and bought in huge numbers by the literate, to show the extent to which such general and lay media conveyed medical information. The fact that we do not, in many cases, even know the author of such writings is itself an important clue to the widespread circulation of medical information independently of the medical profession as such, presumably because of a high level of public need and demand. It is, of course, fascinating as well, that much of the medical intelligence presented in these almanacs was directed to animal diseases – indicative of the fact that there were, at that time, far more livestock in Britain than people and of the high monetary value placed upon cattle, sheep and horses: it was after all a society in which, at 'wife-sales', you might exchange your spouse for a bull!

The public circulation of medical knowledge also forms the subject-matter of two contributions which examine the French-speaking Swiss world of the eighteenth century. Both Micheline Louis-Courvoisier and Séverine Pilloud and Michael Stolberg draw heavily upon the rich archives left by the Genevan doctor Samuel Auguste Tissot, a distinguished practitioner to whom thousands of correspondents wrote, seeking medical advice, remedies and sympathy. Most of those letters to him have survived – unlike, alas, his replies to them. From the letters to Tissot, we can judge not merely the sicknesses suffered by the literate classes of the Enlightenment era but, perhaps more importantly, their health anxieties; their feelings towards their bodies in sickness and in health; the stores of formal and informal medical knowledge they possessed; and the nature of the relationships (powerful, authority, deference, patronage?) into which they were prepared to enter with a prestigious doctor.[1]

The pair of articles by Louis-Courvoisier and Pilloud on the one hand, and Stolberg on the other, constitute, if you like, respectively 'form' and 'content'. The former article concentrates upon the formal qualities of the correspondence as such. Such topics are of great interest, because a majority of such letters regarding health written to Tissot were not actually written by the sick person in question but by (for instance) their own physicians, family members, or other respected members of their community, for example, a minister. Through these letters we obtain, in intricately mediated forms, a kaleidoscope of the beliefs and emotions of the sick person, the responses of those near to them, and the views of other authority figures. In such correspondence extraordinary candour runs alongside deep obfuscation, when the topics are as personal and embarrassing as habitual masturbation, addictive behaviour or other self-inflicted sicknesses.

The question of the content rather than the form of these letters is more fully addressed by Michael Stolberg. His chapter focuses on the ways in which medical knowledge circulated amongst the laity – that is to say, the constitution of medical authority in the eighteenth-century public sphere, looking at the push and pull effects. He explores such matters as supply and demand, the production and consumption of medical information. And his conclusions are suitably modulated. He notes, for example, how rapidly and successfully the new notion of nervous disorders was taken up, not least by correspondents with Tissot, despite the fact that mechanical models of the body as such were only slowly assimilated, the old language of the humours remaining most popular. Clearly, it was soon accepted that there was something highly eligible about being thought to be nervous.

On the other hand, other items of new medical doctrine seemed to have met with considerable resistance amongst the sick. For example, the new medical orthodoxy that menstruation, as the discharge of waste products, served no particular health-conferring purpose – and for that reason menopause was not something about which to be anxious – seems to have met a rather cold and mistrustful response amongst women. They appear to have clung to the older belief that the menses, as the discharge of bad poisoned blood from the body, were to be welcomed – surely they were also relieved to find evidence of the fact that they were not pregnant yet again! Despite the new medical orthodoxies, menopause remained a cause of alarm – as we would see it, for strong and sound social and gendered reasons. Stolberg's analysis thus demonstrates very clearly the to-and-fro balance of intellectual, cultural and ideological power between patients and doctors in that environment of medical patronage that Nicholas Jewson so suggestively, if incompletely, described over twenty years ago.

Similar sorts of materials are examined also in Alfons Zarzozo's account of the eighteenth-century Catalan medical world as glimpsed through private letters. His study brings out the drama of disease and the vividness of the accounts provided by sufferers of their own condition, including their (religiously-primed?) preparedness to contemplate their own likely mortality. Was not this, he suggests, connected to the milieu of violent, unpredictable and dangerous political change in the Revolutionary era? It is clear from Zarzozo's analysis of these letters that medical and political aspects of upheaval, calamity and death mingled in the minds of these sufferers.

It is one of the great virtues of Zarzozo's contribution that he enters into the fevered brains and passions of the correspondents he analyses.

From his and other accounts it is clear that the early modern doctor did not possess in and of himself any automatic, unchallenged authority as physician: rather his ideas and opinions were there to be used – take it, leave it, or modify it – by his patients and the wider cultural circuits.

How it was that practitioners of various kinds obtained some authority and prestige in the first place is expertly diagnosed in Yaarah Bar-On's 'Neighbours and Gossip in Early Modern Gynaecology'. Drawing very heavily on the memoirs of Louise Bourgeois, this exploration demonstrates how medical authority was the product of circulating 'gossip' in fields of discursive force. In circumstances of sickness, it was frequently the testimony of friends and family, neighbours and gossips – both friendly and malicious – which determined the story or verdict which came to be affixed to a particular illness episode or medical intervention. It was frequently a family member or servant whose 'gossip' added a particular vital piece of information which had bearings upon a case. Such a person might also, in retrospect, supply the judgement upon the value (or uselessness) of a doctor. All this is conveyed vividly in Bar-On's reconstruction of many cases which the distinguished midwife Louise Bourgeois retailed respecting her knowledge of the successes and failures of her rivals. Bar-On particularly reconstructs how Louise Bourgeois succeeded in inveigling herself into the court as the midwife of Marie de Medicis, the new wife of Henry IV. It is a fine illustration of the manner in which the court and its circles of patronage and interest constituted the mediating channels for medical – or at least, obstetrical – power in that period.

The nature and foundations of medical expertise and authority are deconstructed in another cluster of chapters. In 'Mediating Sexual Difference: the Medical Understanding of Human Hermaphrodites in Eighteenth-century England', Palmira Fontes da Costa looks primarily at a series of cases in which the display of sexual 'freaks' in London's West End in the mid-eighteenth century – an expression of exploitative commercial entrepreneurship – became the subject of medico-scientific mediation by the Royal Society and its scientific penumbra. In particular, one Fellow, James Parsons, was insistent that there was no such thing as a true hermaphrodite: the so-called 'hermaphrodites' on show – indeed, all possible 'hermaphrodites' – were really only malformed and defective females.

Of course, Parsons's reading reveals certain gross sorts of male biomedical prejudice – a basic assumption that it would naturally be women who were malformed rather than men. But, argues Palmira

Fontes da Costa, it should above all be interpreted as signifying a rather authoritarian insistence upon the unyielding orderliness of nature – there can only be two real sexes – and hence as an attempt at normalization by those claiming medical and scientific authority. Yet the irony is that James Parsons and like-minded medics proved in the end to be rather like the mythical Cnut who strove to stay the waves: the powers of commercial exploitation and popular imagination greatly exceeded the capacity of the medics to compel a particular scientific line.

The limits (there shown) of medical power are revealed very clearly in other chapters as well. Constance Malpas's 'Jules Guérin Makes his Market: the Social Economy of Orthopaedic Medicine in Paris' offers an account – highly entertaining at that – of the activities of a respected but exceptionally entrepreneurial physician in the wider world of nineteenth-century Parisian culture. Guérin was one of those doctors deeply involved in the genesis of orthopaedics, making bold claims for the ability to straighten crooked backs and rectify other deficiencies, and seeking fame and fortune by setting up institutions to those ends. As such, Guérin forms a good instance of attempts to turn enterprises which had long been tarred with the brush of quackery into respectable specialisms. This required energetic rhetorical and propaganda activity in the medical marketplace. Guérin and others were, of course, making a pitch to cash in on the great preoccupation with fitness and beauty (and hence marriageability) in Louis Philippe's France, dominated by the maxim *enrichissez-vous*, and later in the *belle époque*. But the irony of Guérin's career is that, although he attempted to commandeer the media, public culture got its revenge upon him and his like by lampooning and bringing down to earth the claims of him and others to be able to make the crooked straight. That is precisely what Flaubert (himself a doctor's son) did in *Madame Bovary* in his satire upon Charles Bovary's failed orthopaedic intervention in the case of the lad with the club foot. Here the pretensions of the doctors were yet again laughed out of court by a public suspicious as ever of medical claims – if willing to the point of gullibility to embrace them whenever possible.

Contemporaneously with Guérin is the story told by Logie Barrow of vaccination in England. 'Clashing Knowledge-claims in Nineteenth-century English Vaccination' narrates an intricate, sometimes tragic, yet also hilarious story of the consequences of gung-ho doctors promoting smallpox vaccination almost as though it were the True Faith, and the secondary consequences of the state taking the vaccinationists at their word when it legislated in the 1850s for compulsory vaccination. Through one of history's many ironies a medical procedure became one of the great articles of faith in a Victorian state supposedly dedicated to liberty and individualism. It set itself up to be targeted and destroyed not only by those who believed that smallpox vaccination was an error but also by those who hated the arrogant authority of the doctors, the politicians and the state; who resented encroachments upon civil liberties; who believed that the individual body was sacrosanct, as were the family and parental authority. And so – and it is a story we have seen repeated in scores of cases during our own lifetimes – the fate of a mere medical procedure became utterly entangled in wider webs of political debates. Do science and medicine carry their own authority, or must they look to higher authority – to the state, even maybe to God – in order to entrench themselves?

A sense of scholarly outrage against the effrontery of doctors setting themselves up as gods, authorized to dictate behaviour and judge popular morals, emerges from many recent writings about the early history of sexology. Radical feminist authors (especially those of a separatist lesbian tendency) and gay activists have in particular criticized such writings for their 'normativity' and their alleged advocacy of 'compulsory heterosexuality' pleasures. It has been argued that the agenda of twentieth-century sexology to co-opt women into self-proclaimedly progressive and liberationist movements designed to enhance sexual fulfilment was a means of shoring up the bourgeois family and giving males their oats.

A dissenting note is introduced, however, by Hera Cook. In her 'Sex and the Doctors: the Medicalization of Sexuality as a Two-way Process in Early to Mid-Twentieth-century Britain', Cook counter-argues that the sex experts never possessed much authority themselves – they had no power to make anyone do anything. They were not so much trying to dictate to their readers than attempting to make sense of their own often unhappy and floundering sex lives. Their advice was liberal and permissive rather than authoritarian and coercive. Nobody was forced to read their books or follow their practices. The fact, however, that there was such a huge market in England in the period for such books of sexual and marital advice and birth control information is eloquent evidence, she argues, of the depths of ignorance, misery and sexual failure amongst the British public at this time, and the hunger for an escape. Here is a case of common interest between the medical writers and the reading public.

The idea of 'medical miracles' and 'miracle drugs' is one which the new, critical historiography of medicine was keen to knock on the head. As Toine Pieters shows in his 'Hailing a Miracle Drug: the Interferon', this was a classic example of such a touted pharmaceutical breakthough: it was supposed to be the panacea which would put paid to cancer. The hype came, Pieters shows, not only from a new breed of entrepreneurial biomedical expert but also from philanthropists, from the media, and from various other interested parties. When interferon failed to deliver the expected goods, its management slipped out of the hands of the original researchers and became a public football. New strategies had to be invented – new forms of publicity or damage limitation. Here was a case of literal mediation, i.e. the taking over of an item of medicine by the media.

We may be inclined to think of this as a peculiarly contemporary development. But, in what is in some ways for me the most eye-opening contribution to this book, Catrien Santing's article *'De Affectibus Cordis et Palpitatione*: Secrets of the Heart in Counter-Reformation Italy' demonstrates how medicine and the media were already utterly intertwined back in the Renaissance. Doctors were widely called in, she shows, to examine and authenticate the 'miracles' attending the bodily organs, notably hearts of leading Catholic bishops, cardinals and popes. (Saint) Philip Neri, who had a heart condition which was seen to be wonderful, was centrally involved in the cult of the Sacred Heart of Jesus, a cult launched at roughly the same time as the publication of William Harvey's work on the circulation of the blood.

Anatomists and physicians, such as Andrea Cesalpino, were then called in to remove the hearts of recently deceased high-ranking Churchmen. These hearts, in some cases, continued to beat: a miracle! And so, with great historical irony, the preservation of the organs of holy men, by way of Catholic relics, exactly paralleled the process of dissection and the retention in museums of anatomical specimens. The truths of the faith were given medical authentication, and medicine – anatomy in particular – was sanctified by the Church. It was a very different fate to that which befell Galileo.

But that is the point. The history of the mediation of medicine in culture, art and literature, in religion, in politics, in the law and in society is not something that can simply be read off *a priori* as the frightful story of the imposition of medical power or even as simple polarized conflicts (in the manner of the old warfare of science and religion). What the contributions to this book demonstrate so clearly and conclusively are the unpredictable complexities of individual cases; the plurality of forces – cultural, scientific, social, personal – involved in each instance; and the sheer richness of the ways in which medicine has historically been mediated.

## Note

1. For an excellent discussion of medicine-by-post, see Wayne Wild, 'Medicine-by-Post in Eighteenth-Century Britain: the Changing Rhetoric of Illness in Doctor–Patient Correspondence and Literature' (PhD, Faculty of the Graduate School of the Arts and Sciences, Brandeis University, Waltham, Mass., 2000).

# Acknowledgements

The essays in this volume originated from an international conference we held in Amsterdam in 1999 under the joint auspices of the Huizinga Institute, Amsterdam and The Society for the Social History of Medicine, London. Our first debt lies with all those who gave generous financial support: the two organizing bodies, the British Council, London, the Dutch Royal Academy of Sciences, Amsterdam, The Wellcome Trust and The Royal History Society, both London. Paul Koopman's administrative support was superb. Our thanks are due also to our colleagues and friends who generously gave up their weekend to chair sessions and enliven the discussion: Barbara Brookes (University of Ontago, New Zealand), Marijke Gijswijt-Hofstra (University of Amsterdam), Frank Huisman (University of Maastricht) and Eberhard Wolff (Institut für Geschichte der Medizin, Stuttgart). The conference was blessed with an embarrassment of riches. We are indebted to all those who gave papers which have not been able to appear in this book, for all sorts of reasons. One person should, however be singled out for special praise; his presence ensured a consistently high level of debate and his own lecture, the very last of the conference, electrified the assembled participants: the late Roy Porter. Taking his cue from a previous paper on the Calvinist physician Boerhaave who had punctuated his lectures on medicine with references to God raising his top hat each time in deference as he did so, Porter follwed suit. In a passionate appeal to accord the body more respect in medical history he proceeded to salute with his (borrowed) black hat each time he pronounced the magic word 'body'. It was a star performance earning him an ovation. We did not know then that it was one of his last appearances at an international conference. Porter had already promised his paper to another volume but he offered to write what turned out to be much more than a foreword for this book whose publication he has not lived to see. We dedicate it to him to pay back in very small measure his generosity to so many of us.

All chapters in this book have been extensively reworked, expanded and edited to form a cohesive discussion of the role of mediation in medicine. We thank all contributors for their patience with our endless editorial demands and Louise Taylor for rendering Chapters 5, 7 and 12 into elegant English and Jenny Willis for last-minute expert advice.

Our final thanks must go to the anonymous reader whose enthusiastic report – detailed criticism notwithstanding – seemed to make the effort worthwhile and to our editor at Palgrave, Luciana O'Flaherty and her assistant, Daniel Bunyard.

# Notes on Contributors

**Yaarah Bar-On** was in charge of Jewish Studies and Democratic Education, Israeli Ministry of Education. She is currently deputy director, Diaspora Museum, Tel Aviv University. Her first book on the Social History of Gynaecology in Early Modern France was published in 2000 in Hebrew and will be published in 2003 in English (Rochester Press). Her current project is about the history/story of her grandmother, an Ivy League American pioneer in a communist colony in Palestine (1920–36).

**Logie Barrow** has been, since 1980, Professor of British Social History and Political Traditions at the University of Bremen, Germany. He has been interested for over three decades in democratic and elitist approaches to knowledge, and is currently writing the history of vaccination in England, 1796–1907. His most recent book is *Democratic Ideas and the British Labour Movement, 1880–1914* (Cambridge University Press, 1996, jointly with Ian Bullock).

**Willem de Blécourt** is a historical anthropologist and a fellow of the Huizinga Institute, Amsterdam. He has published widely, amongst many other subjects, on the field of irregular healers. He co-edited several volumes on the social and cultural history of medicine and his monograph on women healers in the Netherlands, 1850–1940 (*An Army of Amazones*) was published in 1999 in Dutch by the Amsterdam University Press. He is also an expert on the history of witchcraft and is currently engaged in a study on werewolves.

**Hera Cook** is currently a postdoctoral fellow at the University of Sydney, where she is undertaking research into the history of emotion in England from 1930–80. She obtained her DPhil from the University of Sussex. Her book *The Long Sexual Revolution: English Women, Sex and Contraception, 1800–1975* is being published by Oxford University Press in 2003.

**Louise Hill Curth** received her PhD from the University of London (2000) and currently lectures in the Department of History at the University of Exeter. She has published on popular health care for humans and animals in journals such as *The Social History of Medicine, Animals and Society* and *Seventeenth Century* and is the editor of the forthcoming *From Physick to Pharmacology: Five Hundred Years of British Drug Retailing*. She is currently working on a major project on medicine and the popular press in early modern England.

**Palmira Fontes da Costa** is an Assistant Professor in History of Science at the New University of Lisbon. She received her doctorate from the University of Cambridge in 2000 and has published articles on natural history, medicine and the understanding of extraordinary phenomena of nature in the eighteenth century in various journals, such as *Studies in the History and Philosophy of Science* and *Notes and Records of the Royal Society of London*.

**Micheline Louis-Courvoisier** is a historian whose thesis on the everyday life in a Geneva hospital (1750–1820) has been recently published under the title *Soigner et consoler*. She has since spent three years researching eighteenth-century epistolary consultations. For the last two years she was in charge of implementing a medical humanities programme at the Geneva medical school.

**Constance Malpas** is Manager of Digital Initiatives, The New York Academy of Medicine, concerned with the health of the urban poor. She pursued graduate studies in the history of science at Princeton University. Her current research interests include nineteenth-century pathology and concept formation and disciplinary development in the medical sciences. Her article 'Text, Texture and Transparency: Information Technologies in Early 20th-century Pathology', is due this year.

**Toine Pieters** is Assistant Professor of the History of Medicine at the Free University of Amsterdam School of Medicine and Professor of the History of Pharmacy at Groningen University in the Netherlands. He has written on the history of twentieth-century science and medicine and has recently completed a book on the genesis of the interferons. He is currently working on a study of the co-development of drugs and practices in Dutch mental health (1950–2000).

**Séverine Pilloud** works as a historian at the Institut Universitaire Romand d'Histoire de la Médecine et de la Santé Publique in Lausanne, Switzerland. She has written a book on the local history of socioeducational care of the disabled during the nineteenth century, *Julie Hofmann: une vie de combats auprès des exclus*. She has since researched eighteenth-century consultations for a thesis on the philosophy of health from the lay and medical point of view.

**Roy Porter** probably needs no introduction. In 1979 he joined the Academic Unit of the Institute for the History of Medicine at the Wellcome Trust, London where he became the driving force for the social history of medicine. He was, until his retirement in September 2001, Professor of History at University College London. As a true Renaissance man he became expert of such diverse areas as the history of geology, London, eighteenth-century British ideas and society, medicine, madness, quackery, literature and art on which topics (and many others) he published over 200 books and articles.

**Matthew Ramsey** teaches history at Vanderbilt University, in Nashville, Tennessee. He is the author of *Professional and Popular Medicine in France, 1770–1830: the Social World of Medical Practice* (Cambridge University Press, paperback reprint, 2002) and of numerous articles on the social history of modern medicine and public health, especially in France. Current research interests include the development of professional monopoly in French medicine and cultural differences in responses to disease in the eighteenth and nineteenth centuries. Case studies of the latter have dealt with rabies and with remedies from the old pharmacopoeia based on excrement and human body parts, among other topics.

**Catrien Santing** was a staff member of the Instituto Olandese in Rome from 1998 until 2002. Now she teaches medieval history at the University of Groningen, in the Netherlands. Since publishing her major study *Geneeskunde en Humanisme. Een intellectuele biografie van Theodericus Ulsenius* in 1992, she has written a number of articles, in both Dutch and English, on Renaissance medicine and culture. Presently, she is working on a monograph *The Heart of the Matter*, on socio-cultural construction of the dead human body and parts thereof in Rome between 1475 and 1625.

**Michael Stolberg** is Professor of History of Medicine at Munich University, Germany and from next year at the Institute for the History of Medicine, University Würzburg. His research interests are: cultural, social and intellectual history of medicine (16th–18th centuries) and the history of alternative medicine. Major publications include: *Ein Recht auf saubere Luft? Umweltkonflikte am Beginn des Industriezeitalters* (Erlangen, 1994); *Die Cholera im Großherzogtum Toskana. Ängste, Deutungen und Reaktionen im Angesicht einer tödlichen Seuche* (Landsberg, 1995); *Wolken über der Serenissima. Eine kleine Geschichte der Luftverschmutzung in Venedig* (Venice, 1996); *Die Homöopathie im Königreich Bayern (1800–1914)* (Heidelberg, 1999); *Homo patiens. Krankheits- und Körpererfahrung in der Frühen Neuzeit* (Köln/Weimar 2003).

**Cornelie Usborne** is Reader in European History at the University of Surrey Roehampton in London. She has published on family policy and birth control, *The Politics of the Body in Weimar Germany* (London: Macmillan, 1992); on sexuality: 'The New Woman and Generational Conflict: Perceptions of Young Women's Sexual Mores in the Weimar Republic', in M. Roseman (ed.), *Generations in Conflict* (Cambridge University Press, 1995); on crime, with M. Arnot (eds), *Gender & Crime in Modern Europe* (London: UCL, 1995) and social history of medicine. Her monograph on *Cultures of Abortion in Weimar Germany* will be published in 2004.

**Alfons Zarzoso** is a historian of science and medicine and the curator of the Museum of the History of Medicine of Catalonia in Barcelona. His PhD thesis at University Pompeu Fabra, Barcelona, is on the practice of medicine in eighteenth-century Catalonia. He has published articles in Spanish, French and English, for example, 'Health Care and Poor Relief in Eighteenth- and Nineteenth-Century Barcelona and Catalonia', in A. Cunningham and O. P. Grell (eds), *Health Care and Poor Relief in 18th- and 19th-century Southern Europe* (London: Routledge, 2003); 'Nuisances urbaines et santé publique à Barcelone sous l'Ancien Régime', *Cahiers du Centre d'Histoire 'Espaces et Cultures'.*

# 1
## Medicine, Mediation and Meaning

*Willem de Blécourt and Cornelie Usborne*

To comprehend how health, illness and healing were understood in the past the historian has to enquire into the variety of the meanings attached to these terms. The argument of this book is that meaning is produced by mediation. One of the characteristics of meaning is that it is usually contested; hence a good method to study how people have perceived incidents, objects or verbal expressions, is to examine different and conflicting opinions and to distil a particular interpretation by comparing and contrasting it with others. But meaning can also be mediated, that is, divergent opinions can be reconciled with each other to reach some kind of agreement. Here we have opted for an emphasis on consensus rather than discord.

Mediation is in fact often essential in the medical process. As David Harley has put it: 'For a successful healing the healer's story needs to be connected to the patient's story, usually by a process of adjustment and negotiation rather than by the subordination of patient to doctor.'[1] The assertion that doctor and patient need to be, or indeed were, on an equal footing, is an important corrective to the still widespread notion that there is a natural hierarchy in the healing business with the academic practitioner on top and the patient at the bottom. This does not mean that we deny the role of authority and power in medical history where indeed it is rife. For medical historians to succeed in recreating the medical process of the past, however, they must emulate Harley's 'successful healing' and pay heed to mediation.

How is meaning produced? How is it mediated? There are several possible answers to these questions depending on how we look at medicine and how at mediation. The way medicine is referred to in this volume denotes both sickness and healing; yet, it also has a much wider meaning. Medicine is not just a changing body of knowledge

1

which doctors have been accumulating over time. It is as much subject to cultural diversity as anything else, especially when we are concerned with studying it in the past. As Marijke Gijswijt-Hofstra, Hilary Marland and Hans de Waardt have argued, 'historical understanding has much to gain by attempting to identify indigenous classifications and their meanings in specific contexts'. Medicine is not exempt from this; like every development in history it needs to be conceived as embedded in its own time, geographical location, differentiated according to class, gender and religion. The three medical historians quoted above are also right to stress a further dimension of the historian's craft, medical historians included: an awareness of the constraints caused 'by the limits of [the researcher's] own cultural repertoire'.[2] It is all too easy to impose present-day, and therefore anachronistic and reductionist, notions on past responses, therapies and procedures. The contributions of this volume signal the extraordinary complexity of any attempt to probe the different personal and collective meanings of suffering and therapies, of the experience of ill health and recovery from it in earlier periods. This kind of history is relativistic but therefore no less accurate or relevant. It might indeed be more appropriate to use the plural and talk about 'medicines' rather than the singular 'medicine' if this was not so unwieldy. Thus, we shall stick to the habitual use of 'medicine' as a collective noun which can embody a multitude of changing opinions and practices.

We propose to explore the following three (to some extent overlapping) concepts to unravel medicine and its history: the medical market, the medical encounter and the medical process. After this we will discuss the different ways 'mediation' has been employed and show its considerable scope extending far beyond a simple patient–doctor relationship. Subsequently we will argue that an approach combining medicine and mediation may provide new research avenues and new insights into the meaning of disease, distress, misfortune, health, well-being and good luck in the past and present.

In the Anglo-Saxon social history of medicine the concept of 'medical market' is frequently evoked, yet rarely defined.[3] The way it is applied, however, leaves little room for uncertainty about the economic character of its meaning. 'Market' or 'the commercial medical marketplace'[4] denotes a laissez-faire approach, i.e. an unhindered interaction between supply and demand; it is structured by the forces of competition and the struggle for monopoly. Any interference by the authorities leads to the loss of characteristics of the market and sooner or later to its demise. This was indeed the case, it is argued, in the nineteenth century when the influence of the state together with the 'therapeutic revolution' resulted in a virtual monopoly of the medical profession. Thus, the patient's freedom to 'shop around' for healers, advice and remedies was curtailed after the long eighteenth century. Such an assessment, however, overrates both the perceived effect of medical science and state intervention. Moreover, it also presumes a strict boundary between public market and private household and focuses on the former at the cost of the latter. As a corrective to a purely economic interpretation of the medical market, Dutch social historians of medicine have suggested the role of immaterial exchanges and the production of social and symbolic values also be considered. In this way the 'medical market' becomes a heuristic and elastic device that can be shaped according to local, regional or even global circumstances. Thus the market serves different purposes for and means different things to different participants.

From the patient's perspective the medical market offers the proliferation of available healers and cures; the former includes lay practitioners as well those registered by the guild, church or state. Cures can range from, for example, divine intervention and self-medication to patent remedies and holy artefacts. A patient's definition of illness is ultimately decisive. It may have evolved from the available repository of knowledge of and attitudes to illness, or can be based on the way the patient has articulated his or her complaints in conversation with family, neighbours or acquaintances. The original description of ill health and preference for a cure may change throughout the medical process depending on the authority ascribed to the variety of healers consulted. The patient's therapeutic choice is, moreover, hardly ever free from constraints, be they financial, religious, political or geographical. The search for a cure always takes place within the context of power relations.

From the healer's point of view the market is constituted by the number of potential patients who are consulting him or her on account of certain complaints as well as the competition from other healers in his or her field of expertise. This cross-section of the medical market also includes the healer's reputation and popularity, although these may of course be negative. For the providers of cures the definition of illness is also of crucial importance, but it does not have to conform to that of either patients or indeed other healers. Depending on the kind of medical market and the period in which it is situated, other participants over and above patient and healer may be identified, such as apothecaries, insurance agents, nurses or medical officers of health.

The medical encounter denotes an event in which the different participants in the medical market interact with each other. Mary Lindemann has described it as 'a moment and a space of interchange between patient and healer, and of both with a larger public of friends, relatives, acquaintances, and authorities'. She finds that the nuanced meanings of these interchanges can only properly be appreciated when situated within the greater 'settings of community, state and society'.[5] This often requires careful study and familiarity with a wide diversity of sources on the part of the historian. Judicial records can be an excellent source for medical encounters, for example when the defendant was a lay healer; but this presupposes that legal arguments used by the defence are recognized.[6] This is only one example of many and this is not the place to offer an exhaustive list, especially since the encounter itself still offers rich opportunities for analysis. A careful interpretation needs to convey the correct balance of authority and familiarity, of awe and trust. The material setting and its psychological impact need to be evaluated, too: other patients in the waiting room, the layout and contents of the consultation room, the practitioner's manner and appearance, the medical intrusion into the private areas of the patient's body. Mediation, that is verbal and non-verbal communication, are the most important elements of the encounter, since they facilitate the passage from the complaint voiced by the patient to suggestions for a cure. By its very nature the language used in the interaction refers to images and concepts from outside the consultation room, whether clothed in terms of science or religion.[7] Sometimes language is incapable of conveying meaning, as when certain forms of pain cannot be expressed adequately or when very young children are ill who cannot yet voice their distress. Medical encounters can of course take place entirely without a doctor, as is the case with medical self-help or when a lay person is consulted; or even without either doctor or patient, for example when a sufferer's neighbour talks on her or his behalf to a trader in remedies.

While the medical encounter is restricted to one particular event, the medical process constitutes a series of events which together make up a continuum starting with vague sensations of uneasiness and ending at best with recovery or else with long-term disability or, at worst, with death.[8] Whereas the medical encounter is generally healer-oriented, the medical process is about the experiences of the sufferer, or patient. The latter is more difficult to reconstruct and is therefore rarely given proper attention by historians. This reluctance to engage with the patient's history has little to do with a scarcity of relevant source material. This volume shows again the richness of available sources such as letters and diaries. And there are others waiting to be used, like the material collected under the aegis of folklore, even though this usually does not identify individual sufferers. From the academic viewpoint and within existing historiography the pains of single patients are still often considered irrelevant, or if they feature they do so only as a case study deemed representative, or possibly exceptional, for example, within what is regarded a statistically relevant sample organized around a specific medical symptom. There is also a tendency to dismiss as 'anecdotal' evidence of individual cases, if uncorroborated by 'objective' judgements. Roy Porter's laudable initiative, back in 1985, to create a viable space for the patient in medical history,[9] has as yet not sufficiently opened up the field because of the subordinate status the non-powerful individual is accorded in the historical enterprise, especially with the decline of that type of social history which is not preoccupied by class.

The project to reconstruct the medical processes is, moreover, hampered by the prevailing tendency by scholars to insert 'illness' as a third player, next to doctor and patient into medical history.[10] This personification of 'illness' is misleading as it confuses categories of agency and phenomenon. It also presumes a definition of 'illness' which is ahistorical, since it is depicted as autonomous, quite independent of human agency. This obscures entirely the role of the different participants in the medical market in arriving at a diagnosis through competition, co-operation and consensus. In contrast, the concept of medical process is useful precisely because it reveals the complex system of a number of individuals negotiating a way of arriving at an explanation for somebody's distress or disease which might offer the possibility of treatment.

In the context of these medical configurations, the obvious way to consider mediation is in the sense of reconciling or 'bridging differences' between patient and healers.[11] This usually involved a number of important middle persons. Lindemann describes mediation performed by relatives and servants: 'They, rather than the patient, were often the ones physically present at the medical encounter as bearers of tales or glasses of urine.' She underlines again 'that such mediations were given in the larger transactional world people inhabited', and were, in fact, hardly different from other actions people undertook.[12]

The authors of the following chapters, however, use mediation in its most basic sense of processes of transmission. Different channels of communication thus play a major part in this book, not just in their capacity as historical sources but also as the means of information about health and illness from which certain bits of information were selected and others ignored. Methods of transmitting knowledge and opinions about afflictions, suggestions about treatment and future prevention varied over time and place: hagiographies in sixteenth-century Rome (Catrien Santing), obstetric notebooks in seventeenth-century Paris (Yaarah Bar-On), almanacs for animal health care in early modern England (Louise Hill Curth), letters in eighteenth-century Switzerland and Catalania (Micheline Louis-Courvoisier and Séverine Pilloud, Michael Stolberg and Alfons Zarzoso), newspaper advertisements in eighteenth-century London (Palmira Fontes da Costa), caricatures and comedies in nineteenth-century France (Constance Malpas), newspaper reports in nineteenth-century England (Logie Barrow) and twentieth-century Europe and the US (Toine Pieters) and advice literature by twentieth-century British 'sex experts' (Hera Cook).

The various chapters in this volume elucidate that mediation is not predicated on a one-way process or a mere filtering down of medical concepts. The way information was selected and adapted in the past is only the starting point of the historian's journey unravelling multi-layered practices. Fontes da Costa, for instance, writes about 'the complex web of mediations involved in the making and diffusing of medical knowledge' and points to the fact that medical communications were subject to intricate negotiations, involving agreements and disagreements which eventually did not just transmit opinions but shaped them, too. Barrow's analysis of nineteenth-century campaigns against compulsory smallpox vaccination similarly shows how pro-vaccinators' assertions and promises were passionately debated and contested by a mottled crew of regular physicians, parliamentarians, political radicals, feminists, trade unionists and a large number of working-class parents (mostly fathers) with the result that the issue became highly politicized and effected not only a change in the law but also in attitudes towards academic medicine and in the democratic process as a whole.

Yet in much medical historiography, the present volume included, the concept of mediation is also used in more elaborate and abstract ways. Since mediation presupposes communication, it often relates to pathways of dissemination and the emergence of either form or meaning from those. Thus, when Ludmilla Jordanova writes about ideas in 'their capacity to act as mediators', she invests them with the power 'to shape both conscious and unconscious experience'. In a similar vein, she has language 'mediate' between nature and science. That is, Jordanova believes, scientists can study fragments of nature and draw conclusions about these because they use language as an intermediate tool. Therefore, according to her, only when these 'verbal and visual' languages are deconstructed can we begin to understand the meanings of illness and healing.[13] This is a path, however, we would not wish to tread, at least when it leads to the conclusion that the particular forms in which illness is mediated and communicated have to be discarded for something more 'essential' to emerge: the elemental which remains after the distorting influences of the mediating language are removed. This is both impossible and inpracticable; impossible because language is the necessary means to shape observations and experiences; inpracticable because without it the cultural influence would also be filtered out. Moreover, to let language and ideas do the mediating is to ignore the agency of the human actors who do the speaking and thinking.

It appears that the manner in which social historians of medicine use the concept of mediation informs and explains their basic assumptions. For example, Charles Rosenberg's notion of the autonomy of 'disease concepts' can be readily gleaned from his statement that they 'mediate and structure relationships' between the different participants in the medical process, the patients, physicians, family, medical institutions, etc. Peter Wright and Andrew Treacher, to take another example, have medicine mediate 'broader social forces'. This points to their criticism of medicine as an autonomous force without any links to the surrounding society.[14] These two positions appear as not only too extreme but also, like Jordanova's, as obscuring the very process of mediation itself. That is to say, the cultural approach we propagate here assigns priority to the practice of mediating medicine.

In his discussion of the eighteenth-century anti-masturbation campaign below, Stolberg asserts that as medical knowledge was mediated, it also served 'as a powerful mediating agent'. The latter evidently refers to the adagium, mentioned above, of social contructionism which had social, political and moral values and norms mediated through medical writings. A similar construction can be found in Cook's contribution, where she writes that the mid-twentieth-century doctors who compiled works on sexual advice 'were not merely mediating medicine', that is, transmitting medical concepts and insights, 'but mediating their experience as individuals within the sexual culture into medicine'. In Curth's paper seventeenth-century almanacs revealed a close economic and moral link between animal and human care in that in some cases treatments for small cattle were deemed to be equally applicable to servants. Astrology, too, was mediated by veterinary advice: i.e. the best time for castration was when the moon was in Aries, Sagitarius or Capricorn. As Santing points out, sixteenth-century religious, Counter-Reformation zeal was just so mediated within the Roman cardiac research. At the other end of the time-scale of this volume Pieters describes how a late twentieth-century American committee attempted to promote the cancer-conquering drug interferon by referring to an amalgam of 1970s values, consisting of both laboratory science and allusions to the provisions of nature. But where the Roman Catholics four hundred years ago succeeded in stimulating the cult of the heart and initiated research into the circulation of blood, complete with dissections and vivisections, the Americans were unable to mobilize modern religious feelings associated with 'miracle drugs'.

Mediation is thus shown on several occasions as at least a two-way and a multi-layered process and also as an uncertain enterprise. Medicine is not just a vessel for social norms, it itself also contributes to their construction, either by strengthening, weakening or transforming them. Medical campaigns did not always succeed, they also failed and sometimes indeed they did both, depending on the period under consideration. Nineteenth-century English physicians, for example, who threw their weight behind making vaccination against smallpox compulsory had to admit defeat in the face of massive public disobedience.

The authors of the following chapters point to many promising avenues for new research especially when they identify the people involved in mediation: the mediators themselves. They reveal medicine to be much more than the interaction between doctors and patients. In Santing's chapter, Catholic saints, who between them covered the whole scala of human and animal ailments, were pre-eminent mediators between the supplicant and the divine power of healing. Medical historians will be familiar with various mediating roles, such as family members or neighbours who suggested a particular diagnosis, remedies and healers, or with the men who took sick people's urine to the urinologist or piss-gazer. They are also acquainted with the general practitioner who referred a patient to the specialist, indeed the journalist or politician reporting new cures or, as it may be, medical malpractice. Medical mediation, as Bar-On, Pilloud and Courvoisier and Zarzoso show below, is only a fragment in the intricate network people formed with each other. In her diphtych centred on the famous French midwife Bourgeois, Bar-On analyses the intervention by neighbours and maidservants in the relationship between practitioner and patient and also scrutinizes Bourgeois's own 'networking' to secure a position at the French court. Female gatherings at childbirth emerge here as a more significant event than doctors have let us believe – they themselves were often happy to delegate since they could not countenance the tedious waiting time. By bringing parties together, the intermediaries accumulated status, a cultural capital to be hoarded and spent on the appropriate occasion. Malpas demonstrates how this kind of mediation was instrumentalized by, for example, the nineteenth-century French orthopaedist Guérin. He became famous as well as wealthy by mediating 'effectively between the profession and the commercial marketplace'.

Thus it appears that a healer could act as a mediator, too, and indeed facilitate healing by mediating between people but more broadly by 'mediating between realms of existence', that is helping to cope with birth, sex, marriage and death.[15] Whether they were ancient astrologers, spiritist mediums, Catholic priests, local blessers or modern shamans, healers also mediated between this life and an afterlife, or between heaven and earth. Mediating between sufferer and higher powers may have enhanced the skills and status of the healing expert, but it was also likely to have bolstered religious and, indeed, medical beliefs. A similar process may well be observed when we consider the role of present-day specialists and their links to the Valhalla of Science.

Our brief outline of the process of mediation in medicine is necessarily generalized and sketchy and open to amendments and adjustments. What we had in mind was to offer a model, a research tool, as it were, to facilitate our grasp of the ever-growing territory of the past, rather than replace evidence whenever a particular aspect of the past appears murky. Thus we (medical) historians become ourselves mediators between past and present, translators of old languages, habits and customs, and keepers of today's changing historical knowledge. This perspective entails a responsibility towards historical actors and an awareness of the limits of our own cultural baggage. It is bound to enrich our present-day outlook, too.

## Notes

1. David Harley, 'Rhetoric and the Social Construction of Reality', *Social History of Medicine*, 12 (1999), 407–35, esp. p. 423.
2. Marijke Gijswijt-Hofstra, Hilary Marland and Hans de Waardt, 'Demons, Diagnosis, and Disenchantment', in idem (eds), *Illness and Healing Alternatives in Western Europe* (London & New York: Routledge, 1997), pp. 1–13, cit. p. 3.
3. The following is informed by Willem de Blécourt, Frank Huisman and Henk van der Velden, 'De medische markt in Nederland, 1800–1950', *Tijdschrift voor Sociale Geschiedenis*, 25 (1999), 361–82.
4. Andrew Wear, *Knowledge and Practice in English Medicine, 1550–1680* (Cambridge: Cambridge University Press, 2000), p. 21.
5. Mary Lindemann, *Health and Healing in Eighteenth-Century Germany* (Baltimore & London: Johns Hopkins University Press, 1996), p. 352.
6. Cf. Cornelie Usborne, '"Gestocktes Blut" oder "verfallen"? Widersprüchliche Redeweisen über unerwünschte Schwangerschaften und deren Abbruch zur Zeit der Weimarer Republik', in Barbara Duden, Jürgen Schlumbohm and Patrice Veit (eds), *Geschichte des Ungeborenen. Zur Erfahrungs- und Wissenschaftsgeschichte der Schwangerschaft, 17–20 Jahrhundert* (Göttingen: Vandenhoeck & Ruprecht, 2002), pp. 293–326; idem, 'Female Voices in Male Courtrooms: Abortion Trials in Weimar Germany', in John Woodward and Robert Jütte, *Coping with Sickness*, vol. 3, *Medicine, Law and Human Rights: Historical Perspectives* (Sheffield: EAHMHP, 2000), pp. 91–106.
7. Cf. Willem de Blécourt, 'Prosecution and Popularity: the Case of the Dutch Sequah', in Woodward & Jütte (eds), *Coping with Sickness*, vol. 3, pp. 75–89.
8. Cf. Flurin Condrau, *Lungenheilanstalt und Patientenschicksal. Sozialgeschichte der Tuberkulose in Deutchland und England im späten 19. und frühen 20. Jahrhundert* (Göttingen: Vandenhoek & Ruprecht, 2000), pp. 214–72.
9. Roy Porter, 'The Patient's View: Doing Medical History from Below', *Theory and Society*, 14 (1985), 175–98.
10. See, e.g., Charles Rosenberg, 'Framing Disease: Illness, Society, and History', in Charles Rosenberg, *Explaining Epidemics and Other Studies in the History of Medicine* (Cambridge: Cambridge University Press, 1992), pp. 305–18, esp. p. 314.
11. Cf. Raymond Williams, *Keywords: a Vocabulary of Culture and Society* (London: Flamingo, 1983), pp. 204–7.
12. Lindemann, *Health and Healing*, p. 358.
13. Ludmilla Jordanova, 'The Social Construction of Medical Knowledge', *Social History of Medicine*, 8 (1995), 377, 363, 375.
14. Peter Wright and Andrew Treacher, 'Introduction', in idem (eds), *The Problem of Medical Knowledge: Examining the Social Construction of Medicine* (Edinburgh: Edinburgh University Press, 1982) pp. 1–22, cit. 13.
15. Cf. Willem de Blécourt, *Het Amazonenleger. Irreguliere genezeressen in Nederland, 1850–1930* (Amsterdam: Amsterdam University Press, 1999), p. 153.

# 2

# *De Affectibus Cordis et Palpitatione:* Secrets of the Heart in Counter-Reformation Italy*

*Catrien Santing*

'Where your treasure is, there will be your heart' (Luke, 12: 34). This text was cited by St Anthony of Padua while performing the funeral rites for a man notorious for his stinginess. Afterwards a surgeon received the order to perform a post-mortem examination, which revealed the disappearance of the heart. It goes without saying that St Anthony knew where it could be found: in the casket where all the other precious objects of the defunct were stored. Pietro Damini (1592–1632) represented the scene on a large panel for the Paduan church of St Canciano. On this, an anatomist is shown opening up the chest surrounded by various onlookers. These include noblemen, saints and monks, but also solemnly clad men in black, who may be doctors of Padua's medical faculty. Although Damini faithfully followed the story of one of Anthony's most famous miracles, in his representation he combined saintly veneration with contemporary interest in anatomy. The dead body does not, for example, rest on a bier but on an anatomical table.[1] The scene cannot have been uncommon in Padua, where illustrious anatomists such as Fabrizio d' Acquapendente and Giulio Casseri performed sections in the anatomical theatre at the Palazzo del Bo.

Hearts, whether of Jesus, of saints or of genuine believers, developed during the later Middle Ages into essential religious instruments and symbols, whereas at the same time they functioned as literal subjects of devotional utterance. St Anthony of Padua (c. 1190–1225), par excellence, became associated with the enflamed heart, the symbol of the love for God that resides in it. Ultimately, this devotion culminated in the devotion of the Sacred Heart of Jesus Christ that was granted an official character in the liturgy only in 1856. In a very literal sense, the hearts of (aspiring) saints could become important – a circumstance

11

which made them immensely desirable for admirers. When in 1308 Chiara of Montefalco died and her body did not fall into decay, her fellow nuns cut open her body, lifted out the entrails and stored the heart in a box. After a while the sisters decided also to open up the still intact organ, which proved to contain a 'cross' inside.[2]

Many similar operations occurred in which saintly bodies were cut into pieces to be divided amongst as many devotees as possible. This practice seems to have flourished especially during the sixteenth and early seventeenth centuries when the Counter-Reformation stressed the bonds between body and soul, envisaging particularly the heart as the vehicle as well as the engine of religious ardour. The visionary Caterina de' Ricci (1531–90), for example, was alleged to have swallowed blood from the wound in Christ's breast. Each time this mystical experience befell her, her own heart filled with blood and almost burst open.[3] A similar thing happened to St Angelo del Pas. On 25 April 1596, as he lay dying in the Roman convent of St Pietro in Montorio, the brothers heard a loud noise: at exactly that moment the saint's heart was pierced by the lance of Christ and it sprang open. His hagiographers found a very obvious explanation for this phenomenon: after he had lived for several years with stigmata on his hands and feet, on his deathbed the holy man also received the signs of Christ's passion internally.[4]

The taste for saintly hearts, literally burning from religious fervour, caused many macabre scenes. No sooner had saints breathed their last than their hearts turned into a major 'collectable'. When Theresa of Avila, according to her autobiography a long-time cardiac patient, died in 1582, one of the sisters immediately fetched a kitchen knife, crudely carved out the heart and, splattering blood all around the place, took the precious relic to her cell. Miraculously, it kept on beating strongly and it caused several crystal containers to be broken before coming to rest in a reliquary. Amongst Teresa's attributes we find a book and a pigeon, but above all the heart and the arrow, mementos of her so-called *transverberatio* when in 1559 an angel pierced her heart and left her totally immersed in divine love.[5]

The very crude, laical autopsy performed on St Teresa, however, may be judged very exceptional for the sixteenth century, a period in which anatomy developed into the leading medical discipline. The treatment of her corpse greatly differs from the skilfully and professionally executed sections upon the bodies of three contemporary saints, St Ignatius Loyola, St Carlo Borromeo and St Filippo Neri, who were canonised on the same day as she, 12 March 1622. The fact is that the remains of the three males were entrusted to the hands and knives of trained, distinguished and even famous anatomists. St Ignatius's corpse was examined by Realdo Colombo, St Carlo's by Gian Battista Carcano Leone and St Filippo's by Giuseppe Zerla, Angelo Vittorio and Bernardino Castellano. All these physicians belonged to the medical elite of Italy and were authors of anatomical and other medical handbooks. In their capacity of professors of medicine their counsel was much desired, whereas in their scientific research they paid a lot of attention to the heart. All these doctors were devout Christians and, as such, very much a part of typical Counter-Reformation religious circles.

This chapter intends to show that interests were often combined, even almost unrecognizably mixed together and therefore should not be studied separately. The complex of motives and interests will be chiefly treated in the context of sixteenth-century Rome, the city where the Counter-Reformation had its origin, where more than anywhere else the above-mentioned saints tried to exert their power and where most of the doctors to be discussed were resident – for longer or shorter periods. In the Jubilee year 1600, at the end of a heavy year of teaching, medical doctor Andrea Cesalpino, for instance, composed a short ecclesiastical history in which he explained the course of history in medical terms of sickness, cure and health. Thanks to Pope Clement VIII, the Golden Age, in which people had needed neither doctors nor wine as their bodies and souls were strong enough in themselves, could be expected to return on earth. The two dragons, Calvinism and Lutheranism, would disappear completely. By way of conclusion, the author praised the friendship between the pope and Filippo Neri. Incidentally, he acted as personal physician to both men.[6]

The sixteenth century saw major social, religious and scientific changes. The general atmosphere of revival and deepening of both insights and emotions was all about anatomization in fragments and then reintegrating them into a new *corpus*. This mental disposition, which Jonathan Sawday described as an 'autoptic vision', permeated all aspects of life.[7] Everything had to be analysed and reordered: society, religion and especially the way religion was lived and experienced. Last but not least, nature had to be conquered and classified. This development expressed itself especially in the field of medicine. Thanks to the special fields of botany and anatomy, the discipline managed to emancipate itself from natural philosophy and made a great leap forward. Besides (ecclesiastical) history, medicine and philosophy, Andrea Cesalpino was also fascinated by botany. In his *De*

*Plantis. Libri XVI* (1583) the vegetable kingdom is classified from a physiological and anatomical point of view, thus laying the foundation for Linnaeus's system of classification.[8] Many anatomical atlases mapped and presented the human body and its parts as detailed as was then possible. Anatomical interest was wide and thus general, but Roman physicians especially held the heart and the circulation of the blood in great medical esteem. At the same time, the heart was the star figure in philosophy and in the arts as well as in religion. Thanks to the discovery of new manuscripts, the Aristotelian–Galenic controversy on the primacy of the brain versus that of the heart, including the location of the soul, reached a new climax. Last but not least, Baroque art and literature aimed at touching, moving, warming, wounding and most importantly enflaming the heart in order to enable the innate soul to liberate itself and fly in ecstasy towards God.[9]

This observation of the universal renown of the heart in various segments of society and scholarly disciplines suggests that the intentions behind sixteenth-century physicians' searches for hearts might also be of a heterogeneous nature. As humanist scholars they brought recently discovered Greek philosophical and medical texts into use and reached an understanding of this knowledge. As professors of anatomy, they were exponents of the Vesalian revolution and championed new discoveries in the human body based on observation and experimentation. As physicians to cardinals and popes they exploited their abilities rhetorically in order to improve their position at court and this they did with great success. When the papal court gathered for Mass in the Sistine Chapel, each member of the court had an appointed place according to his rank. The place of the so-called *archiater pontificalis* (a title referring to a physician in regular attendance to the pope) was in the first row of the benches to the left, proof of his eminent status. As faithful Christians, finally, they often played active roles in the sect-style Counter-Reformation groupings around religious leaders such as Ignatius Loyola, Carlo Borromeo and Filippo Neri. My aim is to discuss the various incentives behind sixteenth-century medical, in particular anatomical activities. This shall be done mainly on the basis of the life and work of the most important Roman Counter-Reformation doctors and their papal and saintly clients, but the argument will be directed towards the 'Italian discoverer of the circulation of the blood', Andrea Cesalpino, in relation to his most famous patient: St Filippo Neri.

## Sickness and health in Counter-Reformation Rome

Rome was an unhealthy place. The town was dirty and there was a lack of clean water, as most of the former Roman aqueducts had not yet been repaired. The Pontine Marshes caused malaria, which led to the death of many ordinary city dwellers, but also of several cardinals and popes. Moreover, there were also common dangers that made life precarious and created work for all sorts of medical healers. The holy sites attracted crowds of infirm persons, a circumstance that required hospitals with proper attendants, amongst whom also educated physicians were to be found. There existed, however, a more remunerative possibility for aspiring healers: the papal court and the many cardinalate households. Popes were often elevated to St Peter's throne at a fairly advanced age, as, of course, others also wanted to stand a chance. Between 1500 and 1600 the average pope was sixty-one years old on election and his reign lasted no more than six years on average.[10] The permanent threat of poisoning called for medical and pharmaceutical knowledge to be near at hand.

No wonder that during this century no less than 118 physicians were made *archiater pontificalis*. Many doctors who had become famous as anatomists or botanists at the universities of Padua, Pavia or Bologna were, so to speak, headhunted, and received job offers from cardinals and popes. Towards the end of their careers, Realdo Colombo, Archangelo Piccolomini, Bartolomeo Eustachi, Michele Mercati and Andrea Cesalpino sought their fortune in Rome. The position of personal physician was often combined with that of professor of medicine at the Sapienza University. Roman health service was controlled by the Collegium Medicorum Almae Urbis that assumed responsibility for healing activities in professional, moral, juridical and corporate respects, including the selection and examinations of all potential professional healers. Many papal physicians became members or even presidents of this council, which also supplied the *protomedicus* of the papal states.[11]

Although in general they had already published their most fundamental works before they became Romans, in the Eternal City these doctors continued their scientific work and published new *summae* of their findings. For this they required not only books that could be found in the Vatican or in the private libraries of many cardinals, but also human corpses. As the authorities were rarely prepared to make the material remains of criminals available for research purposes, anatomists had to raid churchyards as well as catacombs. With the same anatomical objectives, clinical practice in trauma hospitals, such as Santo Spirito in Sassia or Santa Maria della Consolazione, was very much sought after. The possibility of observing wounds and of occasionally laying hands on an abandoned patient's cadaver was considered more important than the meagre financial rewards a hospital could offer. In this manner Bartolomeo Eustachi, who came to Rome in the entourage of Cardinal Giulio della Rovere around 1549, based his *Opuscula anatomica* on the autopsies he performed as an expert taking part in medico-judicial investigations.[12] Sometimes famous patients considered it an honour to donate their material remains. Realdo Colombo boasts of having performed autopsies on the bodies of the cardinals Gambara, Cibo and Campeggi, in a special last chapter of his *De re anatomica* (1559) on 'things seldom to be seen in anatomy'. All this means that, apart from thirty-one criminals whose bodies were handed over for public anatomy, eight sixteenth-century popes, several cardinals and aspiring saints whose corpses underwent autopsy, anatomical progress was largely made on the dead bodies of poor pilgrims.[13]

In many respects medical treatment and advice was held to be crucial to the survival of the papal states. In times of pontifical illness, hundreds of bulletins were issued. After having bribed the physician for news about symptoms and a prognosis, diplomats informed their employers, upon which ambitious cardinals rushed to the Eternal City to be there on time for the machinations around the succession to start. During the conclave, special doctors were appointed to mind the physical well-being of the competing *papabili*.[14] To quote Richard Palmer, who has written the only modern survey of Renaissance medicine at the papal court: 'The unstable politics of Rome emphasized the importance of the doctor's function: for the ultimate uncertainty in the capital of Catholic Christendom was the health of the popes, on whom all else depended. If anyone could influence the wheel of fortune at the papal court, then surely it was the doctors.'[15] This, however, seems to have been a medical influence for better and worse, if we believe the story about the demise of the one and only Dutch pope, Adrian VI, in 1523. His sober lifestyle had caused unemployment for half of the city. Much relief was felt on his death, when Roman youth mockingly adorned the palazzo of his physician, Giovanni Antracino da Macerata, with a laurel wreath on which they had written 'liberatori patriae S.P.Q.R.' (The Senate and People of Rome to the Liberator of the Fatherland).[16]

Apart from practical reasons to employ doctors, there was also the factor of status symbol. The humanist Paolo Cortesi wrote a guide for cardinals *De cardinalatu*, in which he stated that besides a sumptuous palace and a large household, a proper physician was an indispensable feature.[17] Although medical doctors were fully accepted as actors in the Roman cultural and social scene, their status remained somewhat ambiguous. Baldassare Castiglione saw no need to include a medical doctor among the courtiers of his *Il cortegiano*, as in this book doctors only appear as a laughing stock. In his diplomatic capacity Castiglione once wrote a dispatch about the death of Cardinal Rossi in which he doubted that the prelate had died from poison. More probably, 'the doctors, not understanding his illness, have killed him, as they do many others', he concluded.[18]

Interest in medical knowledge expressed itself mainly on an intellectual level and only in the second place was considered to be useful for prevention, counselling, prognosis and therapy. Perhaps wisely, sixteenth-century people, including popes and cardinals, often seem to have refrained from submitting themselves to medical therapies, relying instead on domestic remedies as well as religious curative methods. Amongst popes, cardinals and future saints we meet the same mixture of a constant dependence on medical assistance on the one hand and a systematic refusal of professional recommendations on the other. Pope Pius IV, for example, declined any medical treatment for his gout and believed exercise to be the only remedy. Contrary to medical advice he did not abstain from making his daily round, even when suffering high fevers. Despite a painful gouty leg and a cataract on his retina he climbed the cupola of St Peter's on 29 March 1561.[19] In defiance of his doctors' judgement that ass's milk was bad for his stomach, the next pope, Pius V, continued to drink it as a cure for kidney and bladder stones. Towards the end of his life this medicine seemed to have lost its effect and his stomach problems became worse. Doctors proposed an operation, which the pope declined. He closed his eyes forever on 1 May 1572. Autopsy confirmed the professional diagnosis as three black stones were found.[20] The second next pope, Sixtus V, was an even worse patient. In the summer of 1580 his doctors were desperate. The papal fever did not go down because the pope continued to eat melon and drink white wine cooled in snow brought down from the Abruzzi. At the same time Sixtus, who came from a learned family of lawyers and physicians, countered his physicians' professional diagnosis with quotations from Hippocrates, Galen and Avicenna.[21] The same kind of learned arguments had been popular with Paul IV, who was already 79 on the day of his election. Convinced that he would outlive his centenary father, and despite bad signs such as comets to be seen above the Vatican Palace, he refused therapies against his dropsy. He rebutted doctors with their own sources and argued that a pilgrimage to the house of the Holy Family at Loreto would be more effective.[22]

Many doctors started under the patronage of a cardinal and then moved on to a pope. Renaissance popes and cardinals, however, did not employ learned doctors for practical reasons only. Most of them cherished genuine literary and philosophical interests and many of them supported the production of major treatises, translations and editions. These books concerned topics such as botany, mineralogy, zoology, medicine and especially anatomy, and they show us that in those days, medical works were fully accepted as a branch of general knowledge and learning. This signifies that even very technical works, such as Archangelo Piccolomini's lectures on the constitution of the human body, found a willing ear at the papal court.[23] After having served Paul IV as a papal archiater, Piccolomini dedicated his *opus magnum* to another patron, Sixtus V. Cardinal Ridolfi was the major patron of Realdo Colombo and offered to pay for seeing his *De re anatomica* through the press.[24] His pupil and successor, Juan Valverde de Amusco, composed a regimen *De animi et corporis sanitate tuenda libellus* (1552) for Cardinal Girolamo Verallo, then *nuntius* or papal legate at the French court. Thus, in the fifteenth and sixteenth centuries, papal and cardinalate patronage actively sought to improve the understanding of ancient learning in broad areas of natural philosophy, which at that time included medicine. The Studium urbis or Sapienza flourished under papal patronage and provided positions for professors of theoretical or practical medicine, anatomy or botany, a good many of whom, as mentioned, doubled their academic role with that of papal physician.

Rome provided fertile soil for the medical renaissance of the sixteenth century, as many of its leading protagonists spent some time there. Central episodes of this renaissance, such as the first publication in Latin of the complete Hippocratic corpus by Marco Fabio Calvo, including hitherto unknown writings on surgery, took place in the Eternal City. This applies also to the full recovery and translation of Galen's principal anatomical study, *De usu partium* (On the usefulness of the parts of the body), a work that formed the main stimulus of the anatomical renaissance. It was published either at the papal court or in its immediate surroundings, the households of the cardinals.[25] Papal physicians showed a great interest, too, in the recovery of ancient Greek medicine and they wanted to share their finds with a Latin-reading public. Physician and poet Ferdinando Balami translated Galen's *De ossibus* in Latin, under commission of his employer Pope Clement VII, while his colleague, Paolo Messeri, physician to Paul III, did the same with the first two books of Aristotle's *De animalibus*.[26] Ambitious cardinals, naturally, imitated this. Realdo Colombo's patron Cardinal Ridolfi put manuscripts of Greek medical works from his own library at the disposal of the former Pisan professor Guido Guidi to translate them into Latin.[27] The group of anatomists who flourished in Rome were fairly critical of Vesalius, but certainly not from the standpoint of unreflected adherence to ancient texts. With Galen's *De usu partium* and Vesalius's *De humani corporis fabrica* (1543) as points of departure, they rather sought to correct both Galen and Vesalius on the basis of their own researches. These doctors searched for true insights into the body's hidden causes and functions by means of rational contemplation, critical observation and practical experimentation, including animal vivisection.

## Saintly hearts in the medical discourse

Counter-Reformation saints influenced Roman society, which they pervaded with a religious zeal based on reformation and renovation. Many Renaissance medical doctors fell under their spell and as a consequence they tried to combine religious and medical interests. Apart from spiritual satisfaction, ties with religious leaders might also be useful for other reasons. Bartolomeo Eustachi, for instance, when accused of the possession of forbidden books, was saved from the Inquisition by Carlo Borromeo.[28] In his turn the doctor dedicated his *De renibus* to the future saint. Eustachi was an expert on veins on which he wrote a large book, comparing the different views of Galen and Vesalius on the subject, illustrating his examples with descriptions of vivisectional experiments and discussing the passage of blood to and from the heart.[29] For his patron, the cardinal Della Rovere, Eustachi wrote a small treatise entitled *De multitudine*, in which he elegantly combined his various interests. Here the keywords were *plethos* (multitude), *plesmone* (satisfaction) and *pleonexia* (excess), giving blood, the human soul and both their faculties a central role in his argument and connecting them directly with the world soul or Holy Spirit.[30]

Since the heart held such a prominent place in the medical and the religious discourse of the sixteenth century doctors might have considered the opportunity to look into the hearts of saints the chance of a lifetime. Several papal physicians who also, often by order of their employers, frequently attended the bedside of Counter-Reformation saints are at the same time the main actors in the process of the discovery of the circulation of the blood, which ended with William Harvey's treatise *De motu cordis* (1628). In 1548 Pope Paul III called Realdo Colombo (1510–59) away from his Pisan cattedra to make him *archiater pontificalis* and professor of medicine at the Sapienza.[31] In the capacity of most famous physician of Rome the honourable task was conferred upon him in 1556 to perform an autopsy on the body of St Ignatius Loyola, who had chosen the burning heart as the symbol of his Societas Iesu.[32] Angelo del Pas, another 'heart saint', later claimed that, in imitation of his namesake, Ignatius of Antioch, the heart of Loyola had also been inscribed with the name of Jesus Christ, a fact not mentioned by Colombo.[33]

Colombo worked closely together with Michelangelo, who was also deeply interested in anatomy and had promised to execute the illustrations for Colombo's *opus magnum*.[34] Regrettably, Michelangelo was too busy and Colombo died too soon. His sons decided to publish *De re anatomica* in 1559, but apart from a frontispiece it appeared without plates. Colombo's goal to outshine his former teacher and subsequent adversary, Andreas Vesalius, had failed. In Book XI of this work he allotted much space to the heart, whose function and structure he described. Its movements are specified and substantiated by anatomical evidence. Thus, for the first time the so-called little or pulmonary circulation or transit of the blood was explained and proven. This meant correcting Galen, who had stated that blood moves through invisible pores in the cardiac septum from the right to the left ventricle. Colombo's teacher Vesalius had already hypothesized that these pores did not exist, but now their absence was experimentally demonstrated by laying open the thorax of a living dog. The motion of the heart was shown, which dilated when the arteries contracted, while they contracted when in their turn the arteries dilated, leading to the conclusion that blood went from the right ventricle of the heart via the lungs and, after having been mixed with air, made its way back by the pulmonary artery to the left ventricle. In a public demonstration Colombo gave to support his argument, he lighted the vessels of the heart, very quickly cut the heart out of the body and released the dog. With this he demonstrated that, contrary to Aristotle's teaching, the voice did not find its origin in the heart as the animal had barked and walked for a few moments. In his *De motu cordis* William Harvey paid tribute to the contributions of his predecessor.[35]

Even Gian Battista Carcano Leone, the Pavia anatomy professor responsible for the post-mortem examination of Carlo Borromeo, cherished a special cardiac interest. In his *De cordis vasorum in foetu unione* he elaborated on the closing vessels of the foetal heart.[36] He was the first to demonstrate that in the heart of a two-month-old baby the foramen had been closed, as the arterial canal contracted in the first months of a baby's life. By showing the manner in which the large vessels of the heart of a foetus communicate with each other, he also contributed to the discovery of the circulation of the blood.

In 1584 Carcano was ordered to perform the section on Carlo Borromeo. The devout doctor considered this to be a tremendous honour and therefore commemorated it immediately in a special treatise.[37] He took out the entrails and embalmed the body, which made it possible to place the saint on a bier for three days. This had a special effect on the hearts of the people of Milan: they crowded round the cadaver, beating their chests as if, out of an unspeakable ardour for Borromeo, their hearts urged to leave their confinement. It was obvious that the saint's heart was immediately held in veneration. Carcano received orders to take it out and divide it up as the whole of Italy was anxious to get a part. In Rome the church of SS. Ambrogio e Carlo al Corso solemnly received a substantial piece on 22 June 1614 (Fig. 1).[38] In his report the physician expressed the emotions that seized him when holding in his left hand the heart that had enlightened so many other hearts and his surgical knife in the right. Invading the heart with a crude iron instrument caused him deep pain, as here he was holding an organ that had emitted the vital spirits and dispersed virtue, even life itself. Borromeo's vital spirits had brought relief upon a whole diocese and had conferred warmth in a cold, icy era, thus motivating many people to alter their life and become new, reformed Christians. After having expressed these comments, Carcano described the outward appearance of the heart, medium in size, but anatomically perfect. During the operation the organ showed its saintliness by increasing in size, as if it suddenly it had turned into a wide sea. In this case, the physician proves to be a perfect exponent of Counter-Reformation rhetoric concerning the heart by almost equating vital spirits with the fire of the Holy Ghost and hailing the heart as the domicile of the human soul. Anatomization of society had progressed so far that the signs of Christ's suffering no longer needed to be literally found in the hearts of saints. Anatomical perfection began to be favoured.[39] Carcano, nonetheless, remained a true anatomist and medical doctor by

Reliquiario col cuore di San Carlo,
Basilica dei SS. Ambrogio e Carlo al Corso, Roma

*Fig. 1*   Reliquary with part of the heart of Saint Carlo Borromeo, church of SS. Ambrogio and Carlo, Rome.

cutting open Borromeo's chest and heart according to surgical rules and describing the heart in Galenic terms as the producer of the vital spirits.

## The palpitations of St Filippo Neri

St Filippo Neri is often hailed as the Roman apostle. Together with other better-known figures, such as Ignatius Loyola and Carlo Borromeo, he constituted a focal point of religious revival in Counter-Reformation Rome. These three religious leaders, who co-operated a lot, favoured a very literal, almost physical religiosity in which enlightenment by the Holy Spirit was a central feature. During their meetings, the Brothers of Neri's Oratory honoured the Holy Spirit as their special protector and in disputations they discussed God's special instruments: fire, which would enflame the heart; faith, bringing about hope; and iron, enforcing obedience.[40] Man's soul had chosen the heart as its abode, and there it was warmed by the Holy Spirit (Fig. 2). There are other examples of this kind of enlightening. Neri recognized immediately the divine glow on the

S. FILIPPO NERI

*Riceve lo Spirito Santo in forma d'un globo di fuoco nelle Catacombe di S. Sebastiano.*

*Fig 2.*   Filippo Neri making contact with the Holy Spirit in the catacombs of San Sebastiano. Illustration from his vita by P. Bacci.

faces of both Loyola and Borromeo, caused by their love of God and mirroring internal perfection.[41] Divine fervour set him almost on fire and therefore he never, even during cold winters, needed a coat and used to open the windows and door of his cell in order to cool off.[42] On his deathbed he prayed accordingly to God: 'Tui amoris in me ignem accende' (Ignite in me the fire of your love).[43]

The saint's ability 'to know the hearts of men and see their hidden thoughts' made him a very popular confessor, a function he even performed to popes and important cardinals and which resulted in his strong influence on the city.[44] His social programme shows an emphasis on charity since service to God through service to the community was one of the characteristics of the Catholic Reformation.[45] In the case of Neri and Rome, this resulted in the reconstruction of old, and the building of new, hospitals for pilgrims, for syphilitics, for orphans, for reclaimed prostitutes and for the poor and sick in general. In order to train his followers' humility, they regularly had to visit hospitals and sweep the floors, make the beds of patients and even empty their chamber pots.[46]

Filippo Neri lived for eighty years, from 1515 until 1595, which were marked by poor health. His heart condition was just one of a whole series of ailments, amongst which fevers figured prominently, of course. One wonders what people really felt about Neri's medical activities. His many second opinions in cases of severe illness, his loudly recorded medical successes where ordinary physicians had failed, not to speak of his resurrections, must have created at least some professional envy. In 1563 Bartolomeo Eustachi together with Hippolito Salviani and Stefano Cerasa attended his sickbed and sadly predicted the saint's immediate death. They were utterly wrong as the patient lived for another thirty years.[47] Every time the saint's prognoses proved to be correct, 'stupefatti i medici', the vita repeatedly and almost maliciously remarks. Relics and alms to monasteries were always effective, whereas medicines tended to make things worse. Sometimes, Neri restricted himself to giving advice of a more general medical nature such as proposing hygienic measures or a change of air. At other times, divine intervention was invoked and Neri cured thaumaturgically by imposing his hand, by clasping the ill against his bosom with the heavily pouncing heart or even stretching out his own body over that of the sick person.[48] Despite all this – famous medical doctors being repeatedly slain by the powers of the saint who worked many recoveries after the failure of their own therapies – mutual bonds continued to be tight and of a friendly nature.

Contrary to medical expectation, healing happened even to doctors themselves. Michele Mercati (1541–93), for instance, had given himself a very bad prognosis, which was confirmed by his father, who also was a doctor. Neri, however, held different views and promised the doctor another decade of life. Indeed, death was held at bay and the patient lived on. Mercati's reputation did not seem to have suffered as he continued to attend the sickbed of Pope Gregory XIII and later on was appointed personal physician to Pope Clement VIII.[49] Nevertheless, medical care of this patient ran into difficulties too; it was Neri and not Mercati who cured the pope of his chiragra.[50] Doctor Giovanni Battista Modio even left his profession thanks to the influence of Neri. After having published a collection of facetious stories on occasion of the carnival of 1554 and a book *Il Tevere* (1556) in which he underlined the connection between fevers and the numerous floods of this river and asked the pope to restore the ancient aqueducts, he converted. Modio became an Oratorian and dedicated himself to edit the canticles of Jacopo da Todi, a deed for which Neri is hailed as a main source of inspiration. In 1556 the doctor fell gravely ill and according to the prognoses of his colleagues, Pietro Antonio Contugi, physician of Pius IV, and Hippolito Salviani, physician of Julius III, the chances of recovery were minimal. Thanks to the prayers of the saint, which this time were accompanied by a levitation (Fig. 3), these predictions again turned out to be wrong. The health of the moribund was restored. God presented him with another five years and Modio never missed a single gathering of the Oratory.[51]

When Neri was enwrapped in ecstasy, his heart enflamed in love of God. By way of his innate heat he was lifted into the air, sometimes up to the height of several palms of the hand off the ground. This happened especially when he celebrated Mass and elevated the Corpus Christi. At such moments his heart trembled ardently as if asking permission to leave the chest and aspired to ascend towards heaven.[52] The palpitations began in 1544 when Neri was praying in the catacombs of St Sebastiano (Fig. 2). The Holy Ghost, disguised as a ball of fire, entered his body by way of the mouth and installed itself in his breast. This *fuoco d'amore* was so warm that he had to stretch himself out on the floor in order to cool off. After having risen to his feet the feeling of great joy remained and his body started to tremble. The moment he touched his breast, he discovered a tumour as big as a fist. For the remaining fifty years of his life, contact with God manifested itself by heavy, painful beatings in the heart. The defective functioning of this organ caused many crises in the saint's health. During his last illnesses he coughed up large quantities of

Hauendo visitato Gio: B.ta Modio moribondo, ritiratosi in una camera a'far p̃
lui Oratione, fu' veduto eleuato in aria, ch con la testa toccaua quasi il solaro,
circondato tutto di splendori e'l'infermo guari. Vit. Vol g. lib. 3. c.1. n. 11 Occ. nel 1556.

*Fig 3.*    Filippo Neri performs a levitation to cure the physician Giovanni B.
Modio. Illustration from the vìta by A. Gallonio.


blood since bodily heat and palpitations had generated too much of it.
The liquid was carefully collected by one of Neri's devotees, the
Oratorian Francesco Zazzara. The physicians desperately prescribed all
sorts of medicines that only worsened the situation. Neri scorned them
for their foolishness: out of love for God the blood wished to leave his
body in order to bring relief to mankind as a whole.[53] In the end, he died
on 26 May 1595, which, one might argue, was not coincidentally the
night after the feast of Corpus Christi.

Andrea Cesalpino (1519–1603) studied in Pisa with Realdo Colombo and in 1592, on the recommendation of his former pupil Michele Mercati, came to Rome to be appointed as physician to Pope Clement VIII and as professor at the Sapienza. During the following year, in his capacity of heart specialist, Cesalpino was called to the sickbed of Neri to make a diagnosis that was later recorded in the canonization process. Careful scrutiny of the chest revealed that the holy patient suffered from heavy, painful beatings of the heart. This occurred every time he entered ecstasy. Subsequently, the heart would fill up with blood and the artery that led the liquid to the lungs expanded to twice its normal size. In the end, the frequent visions caused the fracture of two of his left ribs, the two closest to the heart. The cartilage of the ribs touched the pericardium, which the palpitations caused to inflate and deflate in the manner of organ pipes. This combination of defects had resulted in a tumour at the left side of the rib cage. Although Cesalpino meticulously described Neri's condition in medical terms, he too ultimately judged the heart condition to be miraculous and supernatural. After all, the palpitations only occurred when the saint concentrated on divine matters and diminished when he started to think of profane things. Divine intervention was the only possible explication, otherwise Neri could not have lived into such an old age. It was a miracle, and therefore another reason to consider him a saint.[54]

The outcome of the autopsy on Neri confirmed Cesalpino's conclusions. Right after the saint's death a section was executed. Since the High Middle Ages this was a regular procedure for popes and kings, who had to be eviscerated in order to prevent immediate putrefaction and to enable the public to bid their farewell at a properly laid out corpse.[55] In the case of Neri the work was done by papal surgeon Giuseppe Zerla and local barber Marcantonio del Bello under supervision of Angelo Vittorio, physician to Pope Gregory XIII and *medico di casa* of the Oratorians, and Bernardino Castellano, head physician of the Santo Spirito hospital.[56] After having opened the chest, they saw that the interior was in an impeccable condition and showed no blemish, only the heart seemed to be bigger in size than might be called usual. The pericardium lacked the water normally to be found there, a fact that was attributed to Neri's excessive ardour, *La fiamma dell' Amor divino*. The entrails of breast and abdomen were taken out of the body, stored in a container, covered with sand and buried in the cemetery of Neri's order, the Oratorians. The cadaver did not undergo any special treatment. Bodily gaps were just filled up with myrrh and roses and then the remains were laid in an ordinary casket. The corpse could be saluted during the next three busy days and then it was placed under the floor of the choir of Neri's own Chiesa St Maria Vallicelliana.[57]

In contrast to what had happened to the heart of Carlo Borromeo, Neri's heart just vanished with the other perishable entrails and was never found again. Physicians and devotees, however, quickly realized their pitiful carelessness. Just eight months later Cardinal Agostino Cusano requested relics of his hero, preferably a piece of the heart. The desire of an important cardinal could not be refuted and so the container with the intestines was dug up. Again, Angelo Vittorio was called to use his professional expertise, but did not succeed in finding the heart in the heap of dead organs.

Because of its transference into a more precious casket, in April 1559 the body was examined again. Apart from Andrea Cesalpino, the physicians Antonio Porti and Rodolfo Silvestris gave testimony of this inspection for the church magistrates. Even then, all bodily parts of the saint had retained their flesh and skin. Citing Aristotle who had claimed that the human soul held its domicile in the heart, Cesalpino stressed in front of the court the flawless state of the thorax with its snow-white colour and still natural looking, soft flesh. This could not be explained by natural or artificial proceedings, but had to have its origin in God. Even in Neri's lifeless body the divine vestiges were still visible and therefore it had not decayed.[58] Only the face had suffered somewhat and for that reason it was equipped with a silver mask, based on the cast taken from Neri's face on his deathbed. In 1922, on the occasion of the tercentenary of his canonisation, the last recognition of his body took place. Once more, the witnesses came to the conclusion that breast and abdomen were miraculously well preserved, whereas in the breast even the lines of the wounds of the autopsy were still visible. Regrettably, Cesalpino's diagnosis could not be completely reaffirmed, while the famous loose ribs had vanished.[59]

Three of the doctors involved decided to write about St Filippo Neri and his heart. In 1613 Angelo Vittorio published his *Medica disputatio de palpitatione cordis, fractura costarum, aliisque affectionibus B. Philippi Neri* in which he discussed Neri's complete case history. In 1594 Antonio Porti was asked by Cardinal Federico Borromeo to write down his findings.[60] Both doctors concentrated on the divine origin of the saint's heart condition, that was to be interpreted as an instrument through which God mediated his influence. Yet the professional medical attitude was never abandoned as all relevant ancient and medieval Arabic sources on the heart figure prominently in the text together with specific ailments and therapies for curing them. Surprisingly, this even applies to the supposed supernatural origin of Neri's physical problems. Vittorio reminded his readers that Hippocrates in his *Liber de morbo sacro* had already distinguished between divine and human illnesses. Of course, the palpitations of Neri had no connection with the natural movements of the human heart. Thus, because of its supernatural origin, physicians inevitably had failed in their attempts to cure Neri's heart in a conventional, medico-therapeutical way.[61]

## Cesalpino and the circulation of blood

During his years as a professor of anatomy at the university of Pisa, between 1546 and 1548, Andrea Cesalpino had been Realdo Colombo's most beloved pupil. He had also attended lessons by Guido Guidi, a professor of theoretical medicine at that same university and was taught Aristotelian philosophy by Simone Porzio. In his most famous work, *Questiones peripateticarum* (first edition, Florence, 1569), Cesalpino continued the work of these three teachers, as it is a *summa* about method, organization of knowledge, the principles of the universe, the explication of the world and its movements, and of organic as well as mental phenomena, all displayed under the patronage of Aristotle. Cesalpino pointed to the heart as the propulsive force of the blood. From the veins into the heart through the cardiac valves and then continuing its way through the arteries, feeding the whole body, the blood was pumped around in a continuous, propitious movement. This process was baptized *circulatio* and stated to be proven by dissection. The heart continuously generated the vital spirit in an ongoing nutrition of the bodily parts, whereas the movement of the blood could be checked by the beat of the pulse.[62] The circular motion brought into mind the course of the planets and stars through the cosmos, a correspondence, the author proudly put forward, already noted by Plato, Thomas Aquinas and Leonardo da Vinci. Next to that the importance of the absorption of air by blood in the lungs was stressed, which prevented the overheating feared by Galen. This led to the conclusion that the heart functions as the central organ for all vessels and all blood, and even for the nerves. In stating this Cesalpino corrected Galen who reserved a leading role for the liver. Thanks to minute anatomical observation and experimentation the differences between veins and arteries were demonstrated, a distinction unknown to Galen. Deservedly or not, even the recent 1999 edition of the *Dizionario Biografico degli Italiani* hails Cesalpino and not William Harvey as the discoverer of the system of circulation.[63]

Andrea Cesalpino referred in very different contexts to the heart of Neri and its problems. In 1602 he published his *Praxis universae artis medicae*, dedicated to his patron Pope Clement VIII. This *Ars medicae* is a typical clinical handbook, as it treats several groups of illnesses according to their location in the body. Book VII, *De morbis thoracis*, discusses the constitution of the heart as well as the ailments specific to this organ, such as palpitations, and therapies to cure them. Cesalpino extensively repeats his views on the heart and the circulation of the blood, carefully making use of all relevant ancient medical sources, especially Aristotle, Hippocrates and Galen. Once more we meet the *continuus motus* making its way from the veins into the heart and from the heart into the arteries.[64] A chapter entitled '*De affectibus cordis et palpitatione*' lists the many disturbances of the heart, carefully distinguished in separate categories. The illnesses of the soul, such as wrath, fear, distress, great joy, desire, love and envy, were reigned by the heart, as Galen had already stated. Since they did not depend on the body, these so-called *perturbationes* were not within the province of medical doctors.[65] The end of this serious, scientifically oriented chapter may seem a major surprise to the modern reader as Cesalpino concludes by repeating the story of the heart complaints of Neri. The account is alike and even the conclusion does not differ from the findings in the canonization process: the palpitations had receded by divine intervention, otherwise the patient would have passed away long before his eightieth birthday. This observation, however, did not prevent Cesalpino including several pages on the treatment of palpitations. When the size of the artery was doubled through natural causes, bloodletting of the right arm was, for instance, the solution.[66]

## Painful relief

The secrets of the heart were crucial to Counter-Reformation Roman medical doctors, but the motives for their quest are inevitably influenced by a mixture of philosophical, scientific and religious concerns. Cesalpino tried to save the concept of the unity of body of soul almost desperately. It was his conviction that all vital functions were steered by the heart, which also accommodated a divine part: the human soul. The soul acted by warmth, a divine principle, and depending on its temperature, was more or less animated. The circulation of the blood was a refuelling process for feeding the body and distributing warmth. Thus, Cesalpino's views lay within the context of the multi-stage cosmological order of the 'Great Chain of Being', as he linked his physical world of mortal specimens with the eternal bodies of the heavens.[67] Despite their 'modern' appearance, Cesalpino's views were far removed from seeing the heart as a mere muscle. During the sixteenth century Galen's experiments were imitated and corrected, but he certainly did not lose influence altogether. Religious beliefs prevented acceptance of the concept of a mechanized human body.

Much more than during the Middle Ages, Counter-Reformation devotion revolved around a truly physical experience of religion, centring especially on the heart and its fuel, human blood. In a certain sense the human body acquired the function of a faith engine. Neri welcomed his illness and felt a direct connection between his physical and his spiritual condition. Oratorian Francesco Zazzara recorded his hero's prayers and one of these, a verse from (the Vulgate's) Psalm 50:12, Neri's 'Cor mundum crea in me Deus, et spiritum rectum innova in visceribus meis' (God, create in me a pure heart, and renew the righteous spirit in my entrails), underlines the bodily basis of that time's devotion.[68] Physical pain was the only means to bring about spiritual salvation, a tendency that found its culmination in the heart complaints of various sixteenth- and seventeenth-century saints. That is exactly what we can read in medical treatises as well as in hagiographic writings, and what can also be found in Baroque paintings and sculpture. As a testimonial and homage to Neri's overflowing heart, Alessandro Algardi in 1640 executed a large marble statue of the saint, which is still on display in the Chiesa Nuova. At his knees, an angel holds up an open book with the text of (the Vulgate's) Psalm 118: 32 'Viam mandatorum [tuorum] cucurri cum dilatasti cor meum' (I shall run the way of your commandments, when you enlarge my heart) that was sung on the saint's feast day.[69]

## Notes

* I would like to thank the Netherlands Institute for Advanced Studies (NIAS) in Wassenaar for kindly offering me a scholarship in 2001 and giving me the opportunity to write this paper.

1. Caterina Furlan and Stefania Mason, 'Scienza e miracoli nella pittura veneta del Seicento', in Sergio Rossi (ed.), *Scienza e miracoli nell' arte del '600. Alle origini della medicina moderna* (Milan: Electa, 1998), pp. 116–33, esp. p. 118.

2. Katharine Park, 'The Criminal and Saintly Body: Autopsy and Dissection in Renaissance Italy', *Renaissance Quarterly* 47 (1994), 1–33, esp. 2–3.

3. Bert Treffers, 'Il cuore malato', in *Scienza e miracoli*, pp. 146–56, esp. pp. 146–7.

4. Bert Treffers, *Een hemel op aarde. Extase in de Romeinse Barok* (Nijmegen: SUN, 1995), pp. 38–9.
5. Treffers, *Hemel op aarde*, pp. 130–5.
6. Andrea Cesalpino, *Historiae ecclesiasticae. Compendium usque ad annum iubilei MDC. Traduzione e presentazione di Luigi Condorelli* (Rome: De Luca, 1985).
7. Jonathan Sawday, *The Body Emblazoned. Dissection and the Human Body in Renaissance Literature* (London: Routledge, 1995), pp. 1–15.
8. Giovanni P. Arcieri, *La circolazione del sangue scoperta da Andrea Cesalpino d' Arezzo* (Milan: Bocca, 1939); English edition John P. Arcieri, *The Circulation of the Blood and Andrea Cesalpino of Arezzo* (New York: Vanni, 1945), pp. 161–5.
9. Bert Treffers was the first to make this point. My article is most indebted to his publications. For a discussion of the role of the heart and the body as 'an engine of creed' in the Italian Baroque, see his *Een hemel op aarde* and 'Het lichaam als geloofsmachine. Beeld, metafoor en werkelijkheid in de Italiaanse Barok', in A.-J. Gelderblom and Harald Hendrix (eds), *De grenzen van het lichaam. Innerlijk en uiterlijk in de Renaissance* (Amsterdam: Amsterdam University Press, 1999), pp. 91–115. For a general survey of the heart in Bible and literature, concentrating on the (early) Middle Ages: Eric Jager, *The Book of the Heart* (Chicago: University of Chicago Press, 2000), and for the history of its representation: Pierre Vinken, *Shape of the Heart* (Amsterdam: Elsevier, 1999).
10. Richard Palmer, 'Medicine at the Papal Court in the Sixteenth Century', in Vivian Nutton (ed.), *Medicine at the Courts of Europe* (London: Routledge, 1990), pp. 49–78.
11. Andrea Carlino, *Books of the Body. Anatomical Ritual and Renaissance Learning* (Chicago: University of Chicago Press, 1999), pp. 70–7.
12. Carlino, *Books of the Body*, pp. 88–9.
13. Carlino, *Books of the Body*, pp. 93–119, discusses extensively the selection of the cadavers and the rituals with which a section was surrounded.
14. Gaetano Marini, *Degli Archiatri Pontifici* (Rome: Stamperia Pagliarini, 1784) 2 vols, mentions the medical delegates during the conclaves separately; these are not automatically the physicians of the former pope.
15. Palmer, 'Medicine at the Papal Court', p. 54.
16. Marini, *Degli archiatri*, pp. 322–3.
17. Palmer, 'Medicine at the Papal Court', p. 51.
18. Palmer, 'Medicine at the Papal Court', p. 58.
19. Ludwig von Pastor, *Geschichte der Päpste seit dem Ausgang des Mittelalters*, 16 vols, (Freiburg im Breisgau: Herder, 1906–31), vol. VII, pp. 72–4, 575.
20. Pastor, *Geschichte*, vol. VIII, pp. 44–5, 612–15.
21. Pastor, *Geschichte*, vol. X, pp. 406–7.
22. Pastor, *Geschichte*, vol. VI, pp. 618–19.
23. Archangelo Piccolomini, *Anatomicae prelectiones* (Rome, 1586).
24. Carlino, *Books of the Body*, p. 63.
25. Nancy Siraisi, 'Life Sciences and Medicine in the Renaissance World', in A. Grafton (ed.), *Rome Reborn: the Vatican Library and Renaissance Culture* (New Haven: Yale University Press, 1993), pp. 169–98, esp. 177–83.
26. Palmer, 'Medicine at the Papal Court', p. 57.

27. Carlino, *Books of the Body*, p. 63, n. 104.
28. M. Muccillo, in *Dizionario Biografico degli Italiani* (Rome: Istituto della Enciclopedia Italiana, 1960–), vol. XLIII, pp. 531–6.
29. Bartholomaeus Eustachius, 'De venae quae azygos graecis dicitur', in *Opuscula anatomica* (Venice, 1564), esp. pp. 282–7.
30. Bartholomaeus Eustachius, *De multitudine* (Rome, 1564).
31. *Dizionario Biografico*, vol. XXVII, pp. 241–4; Jerome Bylebyl in *Dictionary of Scientific Biography*, vol. III (New York: Scribner's, 1970–80), pp. 354–7 and Roger French, *Dissection and Vivisection in the European Renaissance* (Aldershot: Ashgate, 1999), pp. 185–7, 196–200, 202–11.
32. Realdo Colombo, *De re anatomica libri XV* (Rome, 1559), p. 266.
33. Angelo del Pas, *Breve trattato del conoscere et amar 'Iddio* (Rome, 1596). On this: Treffers, *Hemel op aarde*, p. 77. Jager, *Book of the Heart*, pp. 90–3, discusses the bodily scripture of the inscribed heart as a popular motif in hagiography, beginning with the legend of Ignatius of Antioch.
34. Carlino, *Books of the Body*, pp. 62–3.
35. K. Russell, 'The *De re anatomica* of Realdus Colombus', *The Australian and New Zealand Journal of Surgery*, 22 (1953), 508–28; E. D. Coppola, 'The Discovery of Pulmonary Circulation; a New Approach', *Bulletin of the History of Medicine*, 21 (1957), 44–77, esp. 48–59; Jerome J. Bylebyl, 'Disputation and Description in the Renaissance Pulse Controversy', in A. Wear, R. K. French and I. M. Lonie (eds), *The Medical Renaissance of the Sixteenth Century* (Cambridge: Cambridge University Press, 1985), pp. 223–45, esp. 236–7; French, *Dissection*, pp. 207–8, gives an account of the experiment on the dog.
36. Giovanni Battista Carcano Leone, *Anatomici libri duo* (Pavia: 1574), part I. On this A. Scarpa, *Elogio storico di G. Carcano Leone* (Milan: Stamperia reale, 1813).
37. Giovanni Battista Carcano Leone, *Exenterationis cadaveris illustrissimi Cardinalis Borrhomaei Mediolani Archiepiscopi* (Milan, 1584). Cf. M. Aurelio Grattarola, *Successi Maravigliosi della Veneratione di San Carlo* (Milan, 1614), pp. 493–5.
38. Bert Treffers, *Il cuore di San Carlo. Una festa e una orazione nella Roma del Seicento. Oratorio dei Lombardi di SS. Ambrogio e Carlo al Corso* (Rome: Associazione Culturale Shakespeare and Company 2, 1998).
39. This development cannot be called general as in the hearts of several saints the *arma Christi* continued to be found (for example: St Carlo di Sezze); see Treffers, *Hemel op aarde*, p. 128.
40. Louis Ponnelle and Louis Bourdet, *Saint Philippe Néri et la société romaine de son temps (1515–1595)* (Paris: Bloud & Gay, 1929).
41. Antonio Gallonio, *Vita di San Filippo Neri. Pubblicata per la prima volta nel 1601. Edizione critica dell' Oratorio di S. Filippo Neri di Roma*, a celebrazione del IV centenario della morte del Santo, con introduzione e note di Maria Teresa Bonadonna Russo (Rome: Presidenza del Consiglio dei Ministri, 1995), pp. 215–17.
42. Gallonio, *Vita*, pp. 28–9.
43. Gallonio, *Vita*, p. 351.
44. Gallonio, *Vita*, pp. 52–4, 121–3, 126.
45. Pastor, *Geschichte*, vol. IX, 121, 132. Sixteenth-century Rome saw a thorough reformation and renovation of hospitals: John F. D' Amico,

*Renaissance Humanism in Papal Rome: Humanists and Churchmen on the Eve of the Reformation* (Baltimore: Johns Hopkins University Press, 1983), pp. 107–10.

46. Ponelle and Bourdet, *Néri*, p. 171.
47. Gallonio, *Vita*, pp. 134–6, and Ponelle and Bourdet, *Néri*, pp. 209–10.
48. Ponelle and Bourdet, *Néri*, pp. 112–13.
49. Gallonio, *Vita*, pp. 199–200, and Pietro Giacomo Bacci, *Vita di San Filippo Neri. Fondatore della Congrazione dell' Oratorio* (Rome, 1818) p. 220.
50. Gallonio, *Vita*, pp. 292–4.
51. Gallonio, *Vita*, pp. 28, 76, 79, and Ponnelle and Bordet, *Néri*, pp. 152–7.
52. Gallonio, *Vita*, pp. 48–51.
53. Gallonio, *Vita*, p. 295.
54. Giovanni Incisa della Rocchetta and Nello Vian, with the collaboration of P. Carlo Gasbarri D. O. (eds), *Il primo processo per San Filippo Neri nel Codice Vaticano Latino 3798 e in altri esemplari dell' Archivio dell' Oratorio di Roma* (Vatican City: Biblioteca Apostolica Vaticana, 1957–63) 4 vols, vol. I, pp. 235–6.
55. A. Paravicini Bagliani, *Il corpo del Papa* (Turin: Einaudi, 1994) pp. 194–6.
56. Incisa della Rochetta and Vian, *Processo*, vol. II, no. 223, pp. 218–20 (Del Bello), no. 224, pp. 220–2 (Zerla) and vol. IV, p. 207 no. 1234. Information on this is not clear. Gallonio, *Vita*, 310–12 includes Cesalpino, but Angelo Vittorio in his own account, *Medica disputatio de palpitatione cordis, fractura costarum, aliisque affectionibus B. Philippi Neri* (Rome: Typographia Camera Apostolica, 1613), pp. 5–6, states that he was in charge and does not mention Cesalpino. On this: Nancy Siraisi, 'Signs and Evidence: Autopsy and Sanctity in late Sixteenth-Century Italy', in id., *Medicine and the Italian Universities, 1250–1600* (Leiden: Brill, 2001) pp. 364–5.
57. Gallonio, *Vita*, pp. 312–16.
58. Incisa della Rocchetta and Vian, *Processo*, vol. I, pp. 222–3.
59. G. Antonelli, *La conservazione del corpo di S. Filippo Neri con appendice su Andrea Cesalpino scopritore della grande circolazione del sangue* (Rome: Pastet, 1922), esp. pp. 23–4.
60. This report was never published; now Milan Bibliotheca Ambrosiana, Cod. Ambros. G 70. On this Luigi Belloni, *L'aneurisma di S. Filippo Neri nella relazione di Antonio Porti* (Milan: Hoepli, 1950) who gives a transcription.
61. Vittorio, *Disputatio*, pp. 6–18, 21.
62. Cesalpino, *Questiones peripateticarum*, bk V, questio IV, pp. 207–8.
63. A. de Ferrari, in *Dizionario Biografico*, XXIV, pp. 123–5. Also: Karl Mägdefrau in *Dictionary of Scientific Biography*, III, p. 177 and Arcieri, *Cesalpino*. The discovery of both the lesser and the greater circulation gave rise to many patriotic disputes, which will not be drawn into the argumentation of this essay, cf. Roger Bainton, 'The Smaller Circulation: Servetus and Colombo', *Sudhoffs Archiv für Geschichte der Medizin*, 24 (1931), 371–4. The discussion concentrated on whether or not Cesalpino was the inventor of the circulation of the blood. Cf. Walter Pagel, 'The Philosophy of Circles – Cesalpino-Harvey: a Penultimate Assessment', *Journal of the History of Medicine and Allied Sciences*, 12 (1957), 140–57; and Pagel, 'The "Claim" of Cesalpino and the First and Second Edition of his "Peripatetic Questions" ', *History of Science*, 13 (1975), 130–8. Also: Mark Edward Clark, Stephan A. Nimis and

George R. Rochefort, 'Andreas Cesalpino, *Quaestionum peripateticarum*, Libri V, Questio iv, with Translation', *Journal of the History of Medicine and Allied Sciences*, 33 (1978), 185–213.

64. Andrea Cesalpino, *Ars medica. Pars Prima. De morbis universalibus ad sanctissimam patrem Domini Clementem VIII Pont. Max* (Rome, 1602), pp. 471–2.

65. Cesalpino, *Ars medica*, p. 473. Cesalpino elaborated his views on this point, and especially on the problem of demonic possession in his *Daemonum investigatio peripatetica* (Florence, 1580). On this: Mark Edward Clark and Kirk M. Summers, 'Hippocratic Medicine and Aristotelian Science in the *Daemonum investigatio peripatetica* of Andrea Cesalpino', *Bulletin of the History of Medicine*, 69 (1995), 527–41.

66. Cesalpino, *Ars medica*, pp. 476–8.

67. Cf. Pagel, 'Claim of Cesalpino', pp. 131–2.

68. Gallonio, *Vita*, pp. 349–51.

69. Jennifer Montagu, *Alessandro Algardi* (New Haven and London: Yale University Press, 1985), 2 vols, vol. II, no. 75, pp. 380–1, and Olga Melasecchi, 'Nascita e sviluppo dell' iconografia di S. Filippo Neri dal Cinquecento al Settecento', in Roberta Rinaldi and Anna Sabitino (eds), *La regola e la fama. San Filippo Neri e l' arte* (Milan: Electa, 1995), pp. 34–49, esp. p. 39.

# 3
# Neighbours and Gossip in Early Modern Gynaecology

*Yaarah Bar-On*

Love and sexuality, pregnancy and childbirth have always been the focus of extensive attention. In the Middle Ages and in early modern Europe matters of women's health involved many members of the community, in a kind of perpetual symposium on women's bodies. In urban and rural cultures alike, the healing process (including pregnancy and birth) took place within the community, and with constant reference to it. Parents, relatives, friends and servants all took part in a process of mediation between conflicting notions of illness and its potential cure. This was achieved through confrontations, but also through shared knowledge and gossip throughout the processes of diagnosis, healing and convalescence.[1]

## Gossip and the process of healing

Midwives and surgeons did not have an absolute authority during the healing process, and were not even always the most important factor in it. The people around the women played an active role in determining treatment and in dictating the patterns of patient–healer relations. They made suggestions, spread rumours, volunteered useful or harmful information, criticized and encouraged, brought about the dismissal of one healer and the hiring of another, negotiated, and served as intermediaries and messengers. Positive public opinion had a decisive effect on the success of professional healers.[2]

'I have met in Faubourg Saint Germain a certain honest and experienced midwife',[3] writes the famous French royal midwife, Louise Bourgeois, in 1609.[4] She continues:

> She secretly delivered the child of a prostitute. There was no sickness that prostitute did not have: She infected the poor midwife

with syphilis in her right hand. The midwife was not aware of the infection. When red spots appeared on her hand, she continued to deliver babies as usual. In this way she infected 35 households: The husbands caught the disease from their wives and the children from their mothers. The men put the blame on the women and suspected their modesty. The wives blamed their husbands. What misery they suffered! A long time passed until they had discovered the cause of the epidemic; in the meantime honest women have paid the price.

A wise woman, a neighbour of the midwife and of her clients, realised that all the infected women have just given birth, all with the assistance of the same midwife. The neighbour saw that the midwife's hand was bandaged and asked her what was the problem. The midwife said that for many days she has been suffering from incurable blisters. The neighbour told the midwife that she must show the hand to a surgeon. It was found to be syphilis.[5]

A midwife, a prostitute, thirty-five birthing women and husbands, babies and whole households, were driven into a local drama of passion, sickness, fear, suspicion and maybe death. But the neighbourhood crisis of trust was resolved by one of the neighbours: she proved to be an exchange of local knowledge and pieced together various pieces of information. She linked the rumours and the bandaged hand, calculated dates and developed a theory of cause and effect. It was she who dared to interfere in the private affairs of the midwife, asking her directly what she was hiding and sending her to a healer. Through her action she served the community as she had protected numerous families from domestic strife.

The neighbour was supposedly a marginal character in that complex social drama, but in effect she was a key actor – and the real hero, who with a single provocative question, changed the entire plot. The story did have a surprising end. Even Bourgeois, who tells in many other cases that she, as a midwife, hated the 'gossip who ruins the good name of midwives', recognized the importance of the neighbour's role and supported her actions.

The neighbourhood 'gossip' was a veritable 'communal institution'; she played an active role in neighbourhood events, storing memories, collecting stories over the years and passing them on. She was called first to lend a hand during childbirth, helped the midwife, and absorbed information she would turn into 'social capital' after the birth. Feverish dissemination of information took place in the community framework surrounding a woman in labour – fresh rumours and old stories, commentaries and remarks on many and varied subjects were made and heard during the long hours of waiting, before and after birth. Rumours spread among members of a given community, enabling them to discuss the state of their community and their status therein, on an informal, everyday level. Gossip – a social process whereby information is exchanged – shaped and changed the public opinion.[6]

Women, who did not work regularly outside the home, spent a considerable part of their time together, as is the case today as well. They cooked and did the washing, exchanged recipes and remedies, consulted with each other about husbands and children and did handicrafts together, as Louise Bourgeois writes in her autobiography:

> I began to do [...] all sorts of odd jobs, such as sewing, handicrafts, embroidery or making garters, with girls from the neighbourhood in which we lived. I taught them, free of charge, those crafts that I knew.[7]

These relations created communications systems and information networks, through which messages were sent and phenomena compared. A curious neighbour, or 'gossip', could be a powerful figure in her immediate surroundings. She would have known the families for a long time, when her neighbours left their houses, what they were wearing, where they were going, when they became pregnant, and whether the dates matched those of a husband's last trip. She knew who delivered the babies, when and how. She did not always use this information for a positive end. She sometimes created scandals, defamed characters, instigated domestic quarrels, and revealed the shameful secrets of individuals and families.[8]

Relations between women neighbours were intensive and intimate, filled with friendship and co-operation, but also with quarrels and hostility – just like the relationship between the neighbourhood midwife and her patients: secrets and intimate friendship could quickly turn into jealousy and hatred. In such cases, hostile neighbours had many means at their disposal with which to harm one another. Mauquest de la Motte, a French seventeenth-century surgeon and man-midwife from Normandy, recounts how one woman, while visiting the sick, called him and told him that a neighbour of hers 'hadn't had her period in some time', and that 'since it had stopped, she had been taking purgatives'.[9] The woman was of course intimating that her neighbour was trying to induce an abortion. All of the sources show that patients did not have exclusive province over dialogue with their healers. Invited or not, the community interfered.

Madame Boursier du Coudray, a French royal midwife in the eighteenth century,[10] told of a long and difficult pregnancy, which serves to illustrate the extent of community involvement:

> In March, bones began to come out ... we determined that they were the bones of a six-month-old foetus. The doctor asked the woman how long she had been pregnant, and the woman replied that she had never suspected that she was pregnant, because her menstrual cycle had not ceased since the last time she had lain with her husband. Nor was her belly so large, and she had not felt the movements of the foetus. Her breasts had not swelled and she had no milk in them. Generally speaking, she had not experienced any of the discomfort that had characterized her previous pregnancies. A few days later however, she was *reminded* [emphasis by Y.B.] that in May, she had craved mackerel – a craving that could not be satisfied because they were too dear. She was reminded that in that same month, ordinary foods had caused her to feel nausea, and she had suffered from a pain in the heart, that she had suddenly craved foods that she ate only rarely, and that in general she had experienced nausea and weakness – signs of pregnancy.[11]

In her answer, the woman denies that she had been pregnant, and claims that she had suspected nothing. She was not, however, the only person with whom the doctor was speaking. Others interfered, and reminded her of things that perhaps she had preferred to forget. They played an important role in 'remembering'. Those who remembered were undoubtedly people who had been intimately acquainted with her over the course of the previous year. They recalled the story of the mackerel that was too dear at that time of year, and they linked this story to other phenomena. They did not allow the woman to forget, and 'helped' her to tell her story, changing it and adding to it.

For good and for bad, neighbours helped one another in many areas, and knew quite well what was going on in nearby homes. The surgeon la Motte tells of another woman who went into labour while home alone, and had just enough strength to go to the window and call to one of her neighbours.[12] The royal midwife Bourgeois tells the story of a woman who gave birth in the middle of the night, and rather than wake her husband to call the midwife, preferred to wake the neighbours. When complications developed, the midwife sent one of the neighbours to fetch a certain type of hot food usually given to the sick, from another neighbour who was caring for a sick person in her house at the time.[13] Without the ties between neighbours, which enabled them to know who was sick and who could be asked for help, how would they have found hot food in the middle of the night?

Like neighbours, maidservants also served as agents of information in the community. The main differences between the servants and the neighbours themselves pertained to their social position *vis-à-vis* the objects of their gossip. Class distinctions resulted in hostility on the one hand and fear on the other. The servants, who knew all about the activities of their mistresses, had no difficulty in gathering information about them, and in disseminating it when expedient. When the sick person – a woman in childbirth, or members of the family – intentionally withheld information from the healer (out of fear, a shame, sense of impropriety or guilt), maidservants served as a very efficient source of information. When the servants co-operated with the healers, they did not always have their mistresses' best interests at heart. Servants often gave healers embarrassing information – out of spite or lack of discretion.

La Motte tells of a woman who stayed in bed for a few days, complaining of amenorrhoea. She suspected that she might be pregnant, but to her relief it was all over quickly and her period returned in a few days – so *she* told the obstetric practitioner. A maidservant had a different story, however. She said that her mistress had an abortion of a small dead foetus, which had been immediately thrown into the fire.[14] This information was certainly not intended to benefit her mistress. In other cases, the motives are less clear: when a certain friend of Bourgeois ('whom I loved dearly') told her of terrible abdominal pains, without explaining the 'reason', a servant added the following embarrassing details:

> She [the friend] told me with great effort, that in the morning she had experienced a great discharge from her belly, and felt ill, as if something had ruptured in her belly. Then she ran a high fever. She told me that she got into bed, and when she wanted to go about her affairs, she began to get up, but felt weak and fell back into bed, without being able to control her bodily functions. The servant told me, in her presence, that the damage had been caused by her own gluttony: she had eaten on the previous evening, two great servings of ice cream, while at the Hôtel de Conti.[15]

In another story, it was again a maidservant who provided Bourgeois with the information she had missed when she left the house, information the woman in labour did not want to give her: 'The servant who opened the door told me: "My mistress has not yet given birth. As soon as you left, the surgeon called another midwife – one of his friends." '[16] The servant, contrary to her mistress, remained loyal to Bourgeois.

Information of this kind, given to surgeons and midwives, was very helpful. When an accident occurred, or when they arrived after labour had already begun, they wished to know what had happened 'before', in the hours, days and even months prior to the event. To this end, they required the help of observers and participants. Henry Deventer, an eighteenth-century surgeon and a man-midwife, describes a birth that had begun a number of days earlier:

> [The foetus] was alive only yesterday, as witnessed not only by the woman herself, but by all of those present as well ... The midwife, they told me, asked two strong women to pull the arm of the foetus until it was torn from its body, hanging only by a strand of flesh.[17]

Deventer did not make do with reports of those present, but checked for himself. Those present, however, provided him with the information, which later served as the basis for his accusations against the violent midwife.

Jacques Guillemeau, a seventeenth-century French surgeon and a colleague of Bourgeois, tells of one attendant who had abused a client, removing a piece of placenta by force, against the orders of the midwife. The surgeons learned of this from a chambermaid who had been watching her at the time: 'She [the chambermaid] explained to us what had transpired, when we inquired into the cause of death.'[18] François Mauriceau, another famous seventeenth-century surgeon, recounts: 'A Year ago I was called to deliver the placenta of a poor woman in San-Marcel [Paris]. It was three days after she was delivered by one of the midwives from the neighbourhood. She suffered badly. The women who were in the room at the delivery described the events and told me that three days had passed since the midwife failed to deliver the placenta.'[19] The information saved the woman.

This pattern is repeated in many stories and in many variations: the surgeon or the midwife arrives to treat the sick, deliver a baby, or perform an autopsy and tries, with the help of witnesses, to understand the course of events. If the woman giving birth is in a state to do so, she may also take part in the discussion, but there are always others involved. The more supplementary information required by the healers – for example, when they were new in the area and unfamiliar with the patient's family or history of the disease – the greater the role played by observers. In such cases, they acted as intermediaries between the healer and the community with which s/he was unfamiliar.[20]

Lack of discretion and the fact that many women giving birth concealed information for reasons of their own, were common problems. In these cases healers had to obtain information through friendly ties with servants or neighbours. Those were happy to furnish information about the patient's life and medical history, and to describe her social and familial relationships. In the case of a number of such testimonies, it was also possible to cross-check information. The healers knew that they could not rely on the reports of these witnesses, who had their own agendas, jealousies and loves, just as the patient had; thus healers tried to confirm or disprove the observers' claims by examining the patient. They note in their writings that examinations in fact confirmed the stories of the servants, neighbours and other observers. In many cases, however, they had no choice but to rely on non-professionals, who were no less reliable than fellow-professionals or clients.

The immediate contact between healers and 'passive observers' or 'marginal participants' was important for the very reason that relations between healers and the 'main participants', the 'decision-makers' – parents or husbands – were difficult and threatening. Healers wished to emphasize their professional authority and protect themselves against allegations in case anything went wrong. 'Minor players' helped healers build a front against the main participants in the drama. Since healers needed them, they developed *ad hoc* relationships with them, in which 'class' differences were blurred.

Relations with servants, neighbours and other observers, the 'minor players', were simpler than with the decision-makers: the healer's superiority was clear, there was no competition between them, they were neither the ones who paid, nor were they the ones who made the decisions. Through the information they possessed, minor participants thus became active and independent players whose opinions were important and decisive. Since these relations were less binding and more relaxed, it was easier to engage in free and open conversation, and to avoid demonstrations of power and displays of 'authority'.

Observers often functioned not only as 'participants' providing information, but also as advisers and judges. When healers hesitated regarding the correct solution, or when their authority was disputed, they encouraged or criticized them. At moments of crisis, these 'observer-participants' became more active. They gave advice, cited examples, told stories, and sometimes even worked shoulder to shoulder with the official healers. The latter were willing to accept such temporary alliances, as long as they did not threaten their status, and helped them to avoid open confrontation with the 'decision-makers'.

Consider, for example, one of Bourgeois' stories, where she describes the role played by two 'minor figures':

> When I delivered the child of a certain woman, her attendant told me – apropos a story of someone else who suffered from the gout – of her mother, who had suffered from the gout for seven years. Anyone who heard her cries had pity on her. A man, on his way to a place of pilgrimage, stopped to ask what had happened. When he was told, he promised that he would teach them a cure, and that upon his return – 15 days later – the woman would be well. They deliberated gravely, whether to do as he said or not, but the patient pressed them. She was given the remedy and in fact was cured, as the pilgrim had promised ... I wrote out the recipe, beseeched by the attendant in order to prepare it for a certain woman who worked for me ... and I also taught her how to prepare it. I, who had often seen her suffering, saw her many times subsequently, and she never fell ill again. I can promise you that anyone who knew her was amazed by her recovery. Once the hearsay of her recovery got around, a young man who wished to know how she recovered, sought her out. She confessed to him that I had taught her how to prepare the remedy. He sent to me, asking whether this was true. I said that it was true, and I sent him the recipe, so that he would be able to see whether the remedies matched.[21]

While Bourgeois was delivering a child – a matter of hours – she spoke with the woman's attendant and with many other people. Practitioners often represented the attendant, a minor participant, as an inferior and even unreliable figure. Midwives constantly sought to emphasize their superiority over the 'ignorant' attendants, and cast aspersions on their professional abilities.[22] When they spent long hours at the bedside of a woman in labour, however, and they had to work together, close relations developed, which blurred class differences and professional conflicts. Alongside the exchange of information for immediate needs and pitching in at times of adversity, neighbours and servants, attendants and passers-by, like the pilgrim, served to loosen tongues, and acted as long-term agents of information. The chatter and the stories, complaining and laughing, created a common milieu.[23]

In a medical system with so many and varied participants, circumstances can change dramatically depending on the reactions of participants to the appearance of new elements, or the introduction of new information. The accessibility of the information to the community, and the community's influence on its application in the early modern era was much greater than that of a community of clients in our day.

The chance arrival of the pilgrim in Bourgeois' story for example, turned into medical intervention that decisively altered the development of the plot. Although he had no professional standing or authority, the patient took him seriously and accepted his advice. The people who were taking care of the sick discussed and deliberated whether his advice should be followed, but it was the patient who made the decision. She didn't care if the man was a charlatan, amateur apothecary, or graduate of the faculty of medicine. She wanted treatment, and that is what she got. Healers were forced to be more open to the intervention of lay people – to hear their advice and try amateur cures. Chance and surprise were thus common elements in the therapeutic process, and the range of solutions was quite broad. The illness and recovery 'scenario' was therefore more flexible and pluralistic than it is today.

## Gossip and career planning

Bourgeois tells us the remedy for gout was a success. The hearsay brought her new admirers who came to ask for the recipe. In other stories, Bourgeois recounts how her fame spread by word of mouth, in the royal court, in the streets of the city, and throughout the kingdom. Stories such as that of the gout remedy, a difficult birth that was successful, a cure that effected a recovery, spread by word of mouth and acted to the healers' benefit.[24]

Through these circulating stories surgeons and midwives, charlatans, saints and magicians, advertised their skills. Healers who were new to a city went to the 'information centres' and told stories about themselves, relying on the urban 'grapevine' to spread their name. Healers sometimes spread specific and well-directed stories, while at other times, they simply hoped for the best, and spread general stories in an unfocused or even unintentional fashion.[25]

The personal story of Louise Bourgeois, as she told it herself in her guidebook for midwives, gives a wonderful example of the way practi-

tioners used and abused gossip, and spread rumours, in order to shape
the public opinion and to influence the social drama:

> After I had been accepted [as a Parisian midwife], I continued to
> work, and attended a large number of women, the poor and unfor-
> tunate, respectable ladies, young and old, and even princesses. In
> the city, little else was discussed but the pregnancy of the queen
> [Marie de Medici, 1601]. The king had chosen Madame Dupuis
> as a midwife for her, the same midwife who had attended the
> Duchesse – who did not recover from her last birth – and whose
> services had also been retained by the Marquise de Guerchville. I
> had never considered attending the queen in childbirth, although
> I thought that she who would be granted the honour, would be
> truly fortunate.[26]

The duchess 'who had not recovered', was Gabrielle d'Estrées, Henri
IV's favourite mistress until her death in 1599. All of Paris knew the
tragic story of the duchess: after an unofficial relationship of many
years as well as a number of illegitimate children, Gabrielle demanded,
when Henri became king, that he divorce Marguerite de Valois, and
marry her. The divorce went smoothly – the royal union was without
issue. Marrying the mistress was trickier. The king's advisers sought a
young virgin queen, unblemished by court intrigue and unsavoury
relationships, a woman without bastards who might eventually lay
claim to the throne. But the beautiful Gabrielle pressed Henri, and
hero though he was in other areas, could not resist her. The problem
was only resolved upon Gabrielle's death a short time before the
wedding. Official documents state that she died of food poisoning,
having ingested a rotten lemon. Perhaps. The king's advisers were
certainly pleased that the demanding mistress had so conveniently
disappeared from the royal scene.[27]

Bourgeois does not hesitate to implicate the old royal midwife,
Madame Dupuis, in this death, although Gabrielle did not die from
complications during pregnancy or while giving birth. Bourgeois
presumes that such manipulation will, in the public opinion, link
Dupuis to the death of the duchess following a difficult delivery, thus
casting her in a dubious light. Moreover, Bourgeois portrays her as the
embodiment of the old generation, whose time has past. Dupuis
represents more than just a frightened and jealous professional group.
She is a symbol of a terrible era that has come to an end, an age of
religious wars, a dying dynasty, a hated and feared Protestant king, a

childless queen and a fertile mistress who would be queen. That period is over, and the new era boasts a king who changed from Protestant to Catholic, from detested to beloved, and adulterer to husband. The new epoch belongs to the new generation, and Bourgeois was their representative: educated, independent and young, fit to deliver the children of the young, fertile and legal wife of the beloved Catholic king.[28]

Bourgeois provides a detailed account of events following the death of Gabrielle. In January 1601, Henri IV and Marie de' Medici were married. The queen immediately became pregnant, and the king intended to retain the services of his mistress's midwife. The queen's doctors however, knew that Marie was not pleased with Dupuis. One day, they met with a number of other physicians in consultation at the bedside of a certain noblewoman who was very ill. It all began at the patient's instigation. She asked after the queen's health, and one of the royal doctors, DuLaurens, replied that the queen was very well, but he did not leave things at that. Professional ethics and confidentiality were not among the cornerstones of medical practice at the time. Instead, he described his concerns at great length, telling his patient that the queen's physicians were rather concerned:

> They know that the queen is entirely unhappy with the midwife the king has chosen for her, and there is nothing more important in childbirth than a pleasant midwife to whom the patient is favourably disposed. They have therefore decided to search for someone younger, who will better understand her condition, and who will work with Dupuis, if the queen remain adamant.[29]

DuLaurens used the consultation to inform the other doctors as well, and to ask them to recommend midwives of their acquaintance. One of them proposed Bourgeois:

> He told them that I had, in his presence, delivered his daughter's children, whose births had been complicated. Marescot said that he had heard my name mentioned, and Violette said that he didn't know me, but that he had heard only good things about me. Ponson said that he knew me well, and that he could not think of anyone better than me. DuLaurens asked to see me, and Ponson offered to accompany him on the way back. Madame de Thou warmly recommended me.[30]

The two doctors went to see Bourgeois at her home. DuLaurens was favourably impressed, but could not promise her appointment, as it was not up to him. He could recommend her, he said, but only if the queen herself raised the subject.[31]

The queen wanted to get a fresh start in life, with a midwife of her own. The two young women (Bourgeois and the queen) had something in common even before they met. They both wanted to remove Dupuis from their lives, and to erase the unpleasant memory of the past to which she belonged. The alliance the two women of the younger generation were to forge was no coincidence, Bourgeois intimates. It was more than personal interest. It signalled the transition from one generation to the next.

From this point on, Bourgeois describes her feverish efforts to secure the post.[32] In the previous chapter, she claimed she had never thought the job would be offered to her, but since the idea had come up, she did not wait for the grace of God, but ascended the ladder on her own. Bourgeois mobilized all of her personal and professional connections, as well as those of her husband. The detailed description of the 'women's network', and the manner in which she gained access to the queen's innermost circle, through her majesty's friends and ladies-in-waiting, is a rare and unusual account. It provides a unique opportunity to understand how the bourgeoisie gained status and positions at the royal court.

Louise Bourgeois' basic assumption in her race to the top of her profession was that the way to the queen's heart is through her ladies-in-waiting. Bourgeois approached acquaintances and acquaintances of acquaintances, neighbours, friends and former midwives. First, she asked for help from a client of hers who was on friendly terms with the queen. This client, repeating the words of the doctor, explained to her the rules of conduct at the royal court: one must not speak with the queen of another midwife in the presence of the king, who greatly favours Dupuis. One time, Bourgeois' friend recounted, someone dared to complain to the king of a mistake made by Dupuis. He became very angry, and forbade her to speak of the matter with the queen. In this fashion Bourgeois described the difficulties she encountered and the king's opposition to her activities. Behind the scenes, Bourgeois set her position – no less adamant than that of the king – against his. She could not openly challenge the king, but she tried to bring the queen into direct confrontation with him.

Bourgeois knew that she could not get her friends and clients to act directly on her behalf – they were afraid to serve openly as intermediaries with the queen. She therefore encouraged them to be circumspect, through casual conversation about birth or a chance meeting with a midwife. She went through all the ladies at court, from clients to their friends, sisters and sisters-in-law, creating a network of information around the queen, to be released to her bit by bit, here and there, as if by chance. One client/friend initiated just such a 'chance' conversation, during the course of which she casually mentioned 'her midwife', Bourgeois, to the queen.

We can see that royal gossip did not remain secret. The queen's doctors spoke in the presence of other physicians, and even in front of patients, on the intimate conflicts between the king and queen. The doctors failed to be discreet in speaking of their royal client, and their aristocratic patient, de Thou, did not hesitate to pass the information on. In this manner, rumours spread quickly. Bourgeois used the information she received, and consciously and systematically exploited the gossip to her advantage. She encouraged her clients and her friends to speak with the queen, and to report the latest developments back to her. She planned her moves well, constantly employing up-to-date information and assessments provided to her by the women at court, all the while creating a flurry of rumours about herself, in the queen's circle. Below the surface, 'underground' channels of communication were paved. In this conspiracy, a plot of women and doctors acting in the queen's best interest, she exploited the rumours and slander around town against Dupuis. The king kept these whisperings in check, but not for long. The court grapevine was very effective, and Bourgeois used this rumour-mongering to her advantage.

All of this, however, was not enough. Weeks passed, and nothing happened. Then Bourgeois remembered the godfather of her youngest daughter, who was born in 1601.[33] The godfather, de Hailly, was married to the youngest daughter of Conti, a nobleman of Italian descent, who was an intimate of the royal couple, and whose Saint Germain palace served as a rendezvous outside the Louvre palace for the king and queen and their close friends. Bourgeois, who lived nearby, knew that they dined there every week. She decided to act through the godfather, and dropped by, as if paying a courtesy call to Madame de Hailly. The first threads of direct contact with the queen were thus woven. Having established an innocent friendship with the godfather's wife, she asked her to speak with the queen, and to send for her when her majesty came to dinner.[34]

The plan was simple, but the young woman refused to co-operate. Once again the forthright Bourgeois had to have the rules of courtly behaviour explained to her. It is not befitting for a young bride who as yet has no need of a midwife to speak of childbirth with the queen, and to introduce a midwife to her. The rules of conduct and protocol in the presence of the queen are strict and very clear. A woman who has not given birth does not speak of childbirth; a subject is not broached unless the queen has expressed an interest in it first; suggestions are not proffered unless advice has been solicited; and subjects that displease the king are not discussed.

Bourgeois, however, finally made progress in her machinations. One Friday in August, she was summoned to the Conti house. Bourgeois waited six hours, from four o'clock in the afternoon, well into the night. She spent the time in conversation with the other women in the queen's circle: idle chatter, gossip perhaps, but a waste of time – certainly not. Casual conversation was a means of establishing friendships, this time with the queen's closest ladies-in-waiting, and especially with Leonora Galigaï, the queen's childhood friend, 'who liked me a great deal', as Bourgeois writes.[35]

Bourgeois admirably negotiated a complex system. Like any ambitious professional, she understood that she must engage in 'political' activities, wholly unrelated to her professional abilities. She was like the lawyer in Balzac's *Colonel Chabert*, who went out every evening to make new contacts and to foster existing ones.[36] Morning and night Bourgeois went about developing her contacts. Her success was threatened, however, by the Marquise de Guercheville, *dame d'honneur* to the queen, close friend of the king, and client of Dupuis.[37] Her presence precluded a meeting between Bourgeois and the queen. The loyal marquise was appointed lady-in-waiting by the king, along with his own sister, in order to prevent just such undesirable meetings and conversations. Late at night, however, when all had gone to their carriages, and Bourgeois had just about given up, she made one last effort, and asked her new friends to arrange a brief encounter:

> I asked my friend to speak with Madame de Hailly, that she might ask Concini [Leonora Galigaï] to speak with the queen about me, now that everyone, including the king, was in his carriage and could not see what was going on near the queen, who was still in her chair.[38]

After a number of conversations the queen turned to Leonora Galigaï (Concini) and said: 'But what would you have me do? The king does not wish to give me one [a midwife] to my own liking.' Leonora, whose great influence on the queen was common knowledge, insisted:

'Madame, your highness could see her without the king's knowledge. All you have is that old woman whom you do not like.' Bourgeois was summoned to the queen, thanks to Leonora, and the initial contact with her. The queen only gave her a glance, 'the length of an *Pater-Noster*',[39] and departed without saying a word.[40]

The following day, the queen made further inquiries from her friends, about Bourgeois. As she recounts, these friends immediately informed Bourgeois, on the very same day:

> At about one o'clock, she [one of the ladies-in-waiting] took the trouble of calling at my home. She asked to see me, and said: '*courage* Madame Boursier, I have good news for you. When the queen saw me today, she asked me who was that woman who I had shown her yesterday and what she was doing at present. I replied that "she is in town, at her home, awaiting the honour of following your orders." '[41]

At the third stage of Bourgeois' progress, following the doctors' initiative and the networking, came another period of waiting. Bourgeois waited two weeks for an answer, but nothing happened. The battle looked lost. The king had already prepared to take a long journey that would last until the birth, and the queen could not change his decision in his absence. But Bourgeois' days of waiting had not been idle ones. Again she gathered intelligence: 'I have heard', she writes repeatedly, speaking of Dupuis and the queen, the doctors and the king, the ladies-in-waiting who were to attend the queen until the birth, and of other midwives hopeful of replacing Dupuis. Every scrap of information was important. Following her report of all the information she had gathered, she summarized as follows: 'I have said nothing of the fact that I have had the honour of seeing the queen, and I have not spoken of the conversations I have had.'[42] While peeking at the cards of others, Bourgeois carefully concealed her own hand. She told no one of her meetings and conversations, for she knew the power of rumours.

After weeks of anticipation, just before the king left for Calais, came the desired turning point in this drama. The crucial moment is intentionally presented by the author as a private conversation – for the first time – between the king and queen. The entire conversation is in an unusual style – continuous dialogue and direct speech:

> When the king departed, he said to the queen: 'you know where I'll be tomorrow. I see by the grace of God that there is yet time until

the birth. After my departure, you will leave for Fontainebleau, and you will lack nothing ... I do not wish other women to go with you, with the exception of the ladies-in-waiting I have chosen for the birth, in order to prevent jealousy and strife. In addition you will have by your side, your chief physician, DuLaurens, another ordinary physician, and Dupuis.' At this point, the queen began to shake her head from side to side, and said: 'I do not wish Dupuis to serve me.' The king was very surprised: 'Why, my dear? Have you waited for the moment of my departure to tell me that you do not want her? And whom do you want?' [The queen:] 'I saw a young woman at the Hôtel de Conti, hearty and energetic, who attended d'Elbeuf.' 'But how? Who showed her to you, D'Elbeuf?' said the king. 'No, she came to me.' The king did not want to leave under such circumstances. He summoned DuLaurens, and told him what the queen had said. The doctor replied: 'Sire, I know her [Bourgeois] well. She knows a thing or two, and she is the wife of a surgeon. For some time now, everyone has known that the queen is unhappy with the fact that Dupuis will attend her. I consulted with the best doctors in this city [listing their names], and we deliberated as to what we would do if the queen should continue to refuse the services of Dupuis – which midwife would we choose? I did not dare to tell your majesty what we all knew of the wishes of the queen, because we knew that you wanted Dupuis.' The king asked: 'Where did you convene?' [The doctor:] 'At a consultation.' 'That's not enough,' said the king. 'Go find her, and bring me the names of 12 respectable women whom she has attended, so that I might determine whether they were satisfied.'[43]

Immediately thereafter, Bourgeois recounts, the doctor came to her, and repeated the king's command. 'I wrote down the names of the last 13 women I had attended, who live close to our home, and I sent him with one of the servants to six or seven women who had given birth.'[44] One of them was d'Elbeuf, who was also summoned to the king. Bourgeois reports d'Elbeuf's meeting with the king: 'D'Elbeuf told me that she had been approached ... She told me all that had passed between the king and her.'[45] Although Bourgeois names her source here, she does not reveal the information itself. Having met with d'Elbeuf, and heard the doctor's report on the list of Bourgeois' clients, the king agreed to change midwives, and departed.

Bourgeois had succeeded. Direct confrontation had ensued between the king and queen. The three people with whom the king spoke – the queen, the doctor and the client – represent the three active elements in her *underground* story: the court physicians and surgeons, the queen's ladies-in-waiting, and the queen herself. The king – chief speaker in all three conversations – is stunned by the state of affairs suddenly revealed to him. His orders, with which the conversation begins, change on numerous points. The conversations with the doctor and d'Elbeuf, stress the king's amazement and the fact that his authority is undermined. Without exception, they all support Bourgeois, and the doctor emphasizes that he speaks not only for the queen and all of the experts, but for the entire nation, for the good of France.

Bourgeois wanted to portray the king as self-assured, convinced that he is the 'master' and that his word is law but who suddenly discovers otherwise. She wanted to highlight her own ability to orchestrate a complicated series of events in which many took part, under the self-confident king's very nose. There are things which are not the province of the king, Bourgeois gently intimates. When the king stubbornly insists, impervious to women's needs, he will not always prevail. Sometimes he will be soundly surprised, and will be forced to change his mind, since an alternative having secretly developed beneath his feet. The representatives of this hidden network suddenly surfaced, at the right moment, putting their plan into practice. Each had played his/her part in perfect co-ordination, and Bourgeois appeared as the driving force behind the court conspiracy. It is a story without certainty but full of initiative, ingenuity and organizational ability.

In Bourgeois' story rumour and gossip play a decisive role throughout her career. Like anyone who wants to influence events, she collects and spreads stories. In this way, one can determine the public climate, plan a course of action, and fashion 'reality' to serve one's needs. Court intrigues have not gone unnoticed in historical research and popular literature. Much has been written about events 'behind the scenes' at the courts of the French kings and nobles, about the influence of officials, mistresses, ladies-in-waiting, servants, confessors and many others.

Bourgeois' story is unique, however. She observes and engages in this system 'from below'. Her descriptions tell us how practitioners and tradesmen – members of the 'third class' – used the same court system. More than that, they reveal to us the secret or untold story of women – the 'Fourth Class': how women created an *underground* of their own, how they used indirect channels, rivalries, friendships, and relations of dependency between patron and client, to advance their career.[46]

Here comes to an end the summer drama of 1601. A few weeks later, on 20 September 1601, Bourgeois' hands brought hope to the kingdom of France:

> At midnight, the king summoned me to the queen, who was feeling unwell. I slept in the queen's wardrobe, with her chambermaids. When I entered the queen's chamber, the king asked: 'this is the midwife, is it not?' He said to me: 'Come, come, midwife. My wife is feeling unwell. Tell me, is this the birth? She is suffering severe pains.' I assured him that it was. The king then immediately said to the queen: 'My dear, you know, I have already told you many times that all of the princes must be present at the birth. I add, in order to further settle the matter, that [this is done] for your greater glory, and the glory of your son.' ... The queen promised to obey all of his instructions. The king replied that he knows how much she is willing to obey him, but he also knows her shy and modest nature, and therefore felt it important to explain that this is the custom at the birth of every queen's firstborn. The king in fact immediately sent for the princes, before the birth itself had begun. In the meantime, the king joked at their expense: 'If you have never before seen three princes suffering greatly, you are about to do so. When these three even-tempered and very benevolent princes see my wife suffering, they will wish they were far away from here.'
>
> [After 22 hours:] He came toward me, near the queen, embraced her, put his lips to my ear and asked me: 'Midwife, is it a boy?' I told him that it was. [The king:] 'I implore you, do not give me only a moment's happiness – it would kill me.' I then removed the Dauphin's swaddling clothes and showed him that it was a boy. He lifted his eyes to heaven, and put his hands together in prayer, thanking God for his grace. Tears the size of peas streamed down his face.[47]

## Notes

1. For a sociological perspective on the praxis of 'Gossip', see Patricia Meyer Spacks, *Gossip* (Chicago: Chicago University Press, 1985).
2. On the importance of gossip in French culture, see Elizabeth Goldsmith, *Exclusive Conversations: the Art of Interaction in Seventeenth-century France* (Philadelphia: University of Pennsylvania Press, 1988).

3. Louise Bourgeois (Boursier), *Observations diverses sur la stéilité, perte de fruit, fecondité, accouchement et maladies des femmes et enfants nouveaux-nes*, I (Paris: A. Saugrain, 1626), p. 211.

4. The best biographical account of Bourgeois was written 150 years ago: Achile Chereau, *Esquisse Historique sur Louise Bourgeois dite Bourcier, sage-femme de la Reine* (Paris: Felix Malteste, 1852). Cf. Wendy Perkins, *Midwifery and Medicine in Early Modern France: Louise Bourgeois* (Exeter: Exeter University Press, 1996).

5. Bourgeois, *Observations*, pp. 212–14.

6. Cf. David Harley, 'Provincial Midwives in England: Lancashire and Cheshire, 1660–1760', in Hilary Marland (ed.), *The Art of Midwifery: Early Modern Midwives in Europe* (London: Routledge, 1993), pp. 27–48.

7. Bourgeois, *Observations*, pp. 101–3.

8. Stories of witchcraft accusations that start from a quarrel between two neighbour women are numerous. Cf. Lyndal Roper, *The Holy Household: Women and Morals in Reformation Augsburg* (Oxford: Oxford University Press 1989).

9. Jacques Gélis, *Accoucheur de Campagne Sous le Roi-Soleil* (Paris: Imago, 1989), p. 21.

10. On Marguerite Boursier du Coudray, see Nina Gelbart, 'Midwife to a Nation: Mme du Coudray serves France', in Marland, *The Art of Midwifery*, pp. 131–51.

11. Marguerite Boursier du Coudray, *Abrege de l'art des Accouchements* (Paris, 1759), pp. 167–9.

12. Gélis, *Accoucheur*, p. 89.

13. Bourgeois, *Observations*, I, p. 113.

14. Gélis, *Accoucheur*, p. 22.

15. Bourgeois, *Observations*, I, pp. 223–4.

16. Bourgeois, *Observations*, II, pp. 51–2.

17. Henry Deventer, *The Art of Midwifery Improv'd*, II (London: A. Bettesworth, 1728), pp. 47–8.

18. Jacques Guillemeau, *De la grossesse et accouchement des femmes* (Paris: A. Pacard, 1621), pp. 234–5.

19. François Mauriceau, *Les maladies des femmes grosses et accouchées* (Paris: J. Henault, 1668), p. 282.

20. Cf. Mauriceau, *Les maladies*, pp. 159–68; Paul Portal, *La pratique des Accouchemens* (Paris, 1685), p. 68; Guillemeau, *De la grossesse*, pp. 234–5.

21. Bourgeois, *Observations*, I, pp. 58–60.

22. Cf. Bourgeois, *Observations*, I, pp. 134–48, 197; II, p. 47; Portal, *La pratique*, p. 210; Cosmé Viardel, *Observations sur la practique des Accouchemens* (Paris, 1671), pp. 191–6.

23. In the popular *Le Caquet de l'Accouchée* (Paris, 1622), an anonymous Parisian pamphlet, we read of a temporary 'equality' that was created around the woman giving birth. Old and young, rich and poor, married and unmarried, shared conversations.

24. Bourgeois, *Observations*, I, pp. 11–13, 16, 18–19, 26–7, 30, 45–8, 65–6, 106–7, 182, 197–203.

25. Cf. the letters of Gui Patin, a seventeenth-century Parisian doctor: *Lettres de Gui Patin (1630-1648)* (Paris: [no publ.], 1907). Cf. Goldsmith, *Exclusive Conversations*, pp. 80–150.

26. Bourgeois, *Observations*, II, pp. 111–12.
27. Cf. J. P. Babelon, *Henri IV* (Paris: Fayard, 1982) and an interesting account of an eye-witness: Louise Margurite de Lorraine de Conti, *Histoire des Amours de Henri IV* (Paris, 1651).
28. Cf. Robert Mandrou, *Introduction à la France moderne* (Paris: Michel, 1961).
29. Bourgeois, *Observations*, II, pp. 112–13.
30. Bourgeois, *Observations*, II, pp. 113–14.
31. Bourgeois, *Observations*, II, p. 114.
32. Bourgeois, *Observations*, II, pp. 116–25.
33. Chereau, *Esquisse Historique*; Perkins, *Midwifery*, p. 26.
34. Cf. on Marie de Medicis' royal court and very private life: L. Batiffol, *La vie intime de la* Reine (Paris, 1906).
35. Cf. on Leonora Galigaï: M. Carmona, *Marie de Medicis* (Paris: Fayard, 1981) and the biography of Leonora's husband: Hélène Duccini, *Concini: Grandeur et misère du favori de Marie de Médicis*, (Paris: Albin Michel, 1991).
36. Honore de Balzac, *Le Colonel Chabert* (Paris: Garnier, 1966), p. 23.
37. Cf. Tiroux d'Harconville, *Vie de Marie de Medicis* (Paris, 1774).
38. Bourgeois, *Observations*, II, pp. 123–4.
39. The length of a *Pater-Noster* is about one minute.
40. Bourgeois, *Observations*, II, pp. 124–5.
41. Bourgeois, *Observations*, II, p. 125.
42. Bourgeois, *Observations*, II, pp. 126–8.
43. Bourgeois, *Observations*, II, pp. 129–31.
44. Bourgeois, *Observations*, II, p. 131.
45. Bourgeois, *Observations*, II, p. 132.
46. Goldsmith, *Exclusive Conversations*; E. Goldsmith and D. Goodman (eds), *Going Public: Women and Publishing in Early Modern France* (Ithaca & London: Cornell University Press, 1995).
47. Bourgeois, *Observations*, II, pp. 150–64, on the Dauphin's birth.

# 4

# Seventeenth-century English Almanacs: Transmitters of Advice for Sick Animals

*Louise Hill Curth*

The study of medical history in the early twenty-first century is generally divided amongst historians who focus either on human or animal medicine. As Roy Porter has noted, these academics have relatively little contact with each other.[1] In a purely academic sense this is somewhat surprising, given the many similarities between the two disciplines over the centuries. It seems likely that this relationship dates back to the beginnings of medical care, when little differentiation was made between treatments for humans and animals. According to epidemiologist Charles Schwabe, in some areas of the world healers worked on the concept of 'one medicine' for many centuries.[2] This relationship has, of course, varied over time according to social and cultural constructs which define the ways in which disease is understood and treated.

Judging by the content of most medical history books, modern historians do not believe that the term 'one medicine' applies to early modern England. While their works acknowledge that human medicine was based on the Hippocratic and Galenic tradition of health and illness, they tend not to include animal medicine. This suggests that there was no systematic basis to early modern animal medicine, and that the diagnosis and treatment of animals was administered haphazardly by 'dangerous' quacks until the foundation of the London Veterinary College in 1791.[3] A second problem is that many historians are still using outdated research methods based on the 'medical discoveries and elite practitioners' approach of the early twentieth century.[4] Historians of human medicine have long accepted that the absence of medical institutions and 'professional' qualifications do not negate the value of other healers. The same should hold true for animal healers, whose work was based on the same beliefs and practices as those who tended to sick people.

This chapter will argue that there is, in fact, ample evidence that animals and humans shared 'one medicine' in early modern England. It is based on the medical content of almanacs which were cheap, annual publications meant to appeal to readers across the whole socio-cultural spectrum. Although they provided readers with a vast range of information, one of their major purposes was to serve as medical mediators by transforming elite medicine into understandable practical advice. This material will be used to show that there was not only a well-developed system of animal health care, but that it was closely linked to that for humans. The basic principles of Hippocratic and Galenic medicine, commonly portrayed as applying only to humans, also related to animals. Preventative and remedial principles, which were the backbone of this belief system, applied equally to animals, both in the way diseases were diagnosed and treated.

## Almanacs

Almanacs were printed and distributed in November and December for the coming year. Their main purpose was to disseminate information about the movements of the planets and their effects on all living things, generally in the form of an astrological calendar. By the early modern period they included a wide spectrum of astrological facts and advice, although most also covered a host of other information ranging from dates of fairs, historical timelines, advice on writing legal documents or medical advice.

Although almanacs had existed for many centuries in manuscript form, the first printed edition was attributed to Johannes Gutenberg in Germany in 1448. By the 1470s large numbers of almanacs were being printed in various other European countries including France, Italy, Hungary, the Netherlands and Poland. In England, however, the evolution of almanacs was much slower. Until the late sixteenth century, most English almanacs were translations of continental ones.[5] The first almanac was printed in England around 1537 and this form of publication became increasingly popular during the following decades. It was during the seventeenth century, however, that almanacs surpassed all previous levels of popularity with some three to four million distributed over the course of the century.[6]

Bernard Capp has argued in *Astrology and the Popular Press* that almanacs were at the height of their importance between 1640 and 1700. This period began with the breakdown of legal controls over the number of publishers allowed to operate printing presses. The resulting influx of new printers led to an unprecedented outpouring of both astrological and medical books. Their popularity continued to grow during the following decades and by the end of the century printers were regularly producing between 350 000 and 400 000 copies in the last two months of every year.[7]

Most surviving almanacs are divided into two major sections. The core of the first section always consists of a calendar marking upcoming astronomical and astrological events for the coming year. This includes major events such as comets or eclipses, as well as various movements and conjunctions of the planets. The second section of almanacs was generally called the 'prognostication'. This part usually contains material that was not very 'time-sensitive', such as tables of weights and measures, distances between towns or medical information and advice.

Almanacs were generally seen as necessities because of the utilitarian material they contained. However, the type of almanac a reader would purchase depended on a range of factors, such as the author and varying levels of the readers' literacy, wealth and sophistication. For example, some authors also produced vernacular medical books, while others had no medical qualifications. Some were hack writers who appropriated either fictitious names or those of long dead authors. Although most almanacs were printed in London, many targeted specific, regional audiences. These included towns from Dover all the way up to Durham, making almanacs the first printed periodicals with national coverage.

Other titles differentiated themselves by targeting specific occupational groups. Although several titles were written for 'country-men', others focused on chapmen, weavers, seamen, shepherds, farriers or constables. Readers with strong religious convictions, or prejudices, could have chosen to purchase almanacs intended for either a Protestant or Catholic audience.

Although the 'targeted' audience differed between editions, most almanacs contained medical advice for both man and beast, either in the form of seasonal advice or medical recipes for specific ailments. The former were generally found in a section entitled 'the four quarters', describing illnesses linked to predicted 'unseasonable' weather conditions for the winter, spring, and summer or autumn. These generally included detailed descriptions of the movements of the planets and the effects they would have on health for the coming year. For example, although the spring was considered to be a healthful time, an almanac writer might predict a drought that would result in a range of diseases for humans and animals.

There were two main types of medical recipes in almanacs. The first, and more basic, was that referring to simples. As the name suggests, these were medicines consisting of one single ingredient. Other recipes were more complex, often containing a variety of different herbs and other organic or inorganic materials. In both cases, authors would also differentiate between those intended for human or animal use.

The medical beliefs and practices referred to in almanacs were based on an orthodox, traditional mixture of astrological and Galenic beliefs and practices. The movements of the stars and planets were believed to have the power to cause imbalances within living creatures, resulting in illness and disease. These were orchestrated by God, who was the 'chief Gouvernour' of the heavens, and who could choose to cause disease, plagues or epidemics to strike individual humans or animals, whole communities or even nations.[8]

Galenic medicine began to play its role once the movements of the heavens had begun their mission to cause disease. This was based on the theory that all living beings had four fundamental elements, qualities or humours. These needed to stay in balance in order to maintain a state of good health. Once an imbalance in the black bile, yellow bile, phlegm or blood occurred the patient would fall ill. As one seventeenth-century writer explained 'if the four humours exceed or alter the proportion in mans body, then shortly commeth disease and sickness'.[9] Judging by the medical advice in almanacs, the same principles appeared to hold true for animals, as well.

## 'One medicine' for man and beast

Although medical historians have failed to notice the similarities in human and animal medicine in early modern England, there is clearly some debate about this among other academics. T. W. Schillhorn van Veen, for example, has suggested that Western Europe had 'one medicine' up until the period of the Industrial Revolution.[10] This argument is supported by a variety of early modern English sources, including almanacs.

Advice on animal health care in almanacs was generally found under the heading of 'husbandry', or 'agricultural advice'. Most of this information referred simply to 'cattel', although horses were the single animal most mentioned. Unlike modern usage, in the seventeenth century it was a generic term that was freely used for most working animals.[11] One of the few exceptions was the dog, regardless of the type of jobs it carried out and animals regarded solely as pets, such as singing birds. Working animals were generally further delineated into categories of 'greater' or 'lesser' cattle. The first type often included 'the horse, ox, cow, &c'. The latter referred to 'lesser sort of Beastes, as Sheepe, Swine, and Goates: and of Fowles, Geese, Peacocks, Duckes, Pigions, Hennes, Chickins and other poultrie'. Deer, conies (i.e. rabbits) and other 'smaller creatures' were also often included under this title.[12]

I have found that the majority of 'pre-veterinary' advice in almanacs was targeted at 'greater cattel' such as horses and oxen. Guidelines for 'lesser beastes' seem to be relegated to the most propitious times for gelding, shearing and mating sheep. In general, the type of medical information for greater cattel was divided into preventative and reme-dial medicine. The former was based on the idea that it was preferable to try to maintain a state of health, than to try to fight disease. In the early modern period this centred on the Galenic concept of the 'non-naturals', which consisted of air, motion and rest, sleep and waking, things taken in (that is, food and drink), things excreted and the pas-sions and emotions.[13]

The first non-natural of 'air', or the weather, was an important consideration for all forms of living creatures. Advice for humans tended to focus on areas of the country with the cleanest air, or infor-mation on the best locations for building homes. The majority of humans and animals lived in the countryside, which was thought to have cleaner air than towns. However, bad weather conditions could adversely affect the health of both rural and urban dwellers. Unusual weather conditions were thought to cause fever, while wet and windy weather was thought to bring measles and smallpox in humans, or the 'rotting' of sheep hooves.[14]

Eclipses, along with other types of celestial phenomena, were thought to affect weather conditions with dire consequences for man and beast. These covered a range of unpleasant, non-medical topics such as quarrelling and fighting, as well as hot fevers and other distem-pers. Almanac writers advised readers that the best way to protect themselves was to follow a good health regime. By keeping their bodies strong, they would be less likely to fall ill during the eclipse.

Animals, of course, could do little to protect themselves from such events. However, by being supplied with warnings of future occur-rences owners could take some steps to protect their animals. For example, the almanac writer Thomas Fowle warned that the impend-ing eclipse would bring 'great losses and decay in their Estate, Cattels and Treasure; it portends death to the greater sort of Cattel, as the Ox, Cow, Horse, or such like'. Since blood was thought to increase in the body during March, it was a good month for 'Physick and Bleeding ... to prevent further dangers'.[15] Such knowledge gave owners the chance to try to prevent, or at least lessen, the effects of the forthcoming eclipse on their animals' health.

Advance warning also allowed those who cared for animals to help to lessen the effects of comets. It was believed that their shape often indicated both the type of coming misfortune and the kind of creatures it would affect. The big comet in March 1665 was thought to have ushered in the great plague in London. A smaller one forecast for 1693 was predicted to cause 'great detriment and many infirmities to those Cattel most useful for man, as Horses, Oxen, Sheep and Hogs'.[16]

While other astral events were thought to cause war or crop failure, some were specifically linked to health problems in both man and beast. The exact consequences varied, and sometimes were thought only to affect certain forms of 'cattel'. Conjunctions involving Saturn were particularly worrying, as the planet was 'the general significator of Labour Disturbance, Anxiety, Melancholy, Poverty, Treachery, Malice, Disgrace &c'. The situation was worsened when an unfavourable planet was in the fifth house, the house of 'sicknesse, servants, small cattle &c'. The astrologer William Lilly warned that a conjunction of Saturn in the fifth house could result in 'detriment, hinderance, losse, consumption, and destruction of four-footed beasts, both greater and smaller, and especially if those are most useful for man, as the Ox, the Horse, Cow, Asse, the Hog, Sheep, Deer, Conny, &c'.[17]

The second and third non-naturals, of motion, rest and sleep, were also vital parts of a good health regime for humans and animals. It was considered detrimental to one's health to over-heat the body either through 'toyl or pleasure'. Popular husbandry books constantly warned that overwork would directly result in 'pestilence'.[18] It was also necessary to ensure that sufficient time was provided for beasts to rest. Furthermore, animals required a dry, secure place to sleep which acted as 'a medicine to that weariness, as a repairer of that decay' that resulted from hard labour. William Dade warned that: 'When a Beast is weary or tire, it may be by weakness, or poorness of Body, and that surely will bring some disease or the other upon him in a short time.'[19]

Sleep was another important part of the Galenic health regime, and one much discussed in early modern medical works. Both too little or too much sleep were thought to have adverse effects on the health of both humans and animals. The correct amount, however, was believed to 'comforts nature much, refresheth the memory, cheers the spirits, quickens the senses'. Sleeping chambers, or animal shelters were also required to be dry, secure places to rest so that the body could repair itself after a hard day's work.[20]

Dietetics, which was the fourth non-natural, has been called one of the most ancient branches of the therapeutic art. Contrary to its modern definition, in the seventeenth century, this generally referred to the place of diet in a good health regime.[21] This included the type, amount and wholesomeness of the food and drink consumed. While humans could monitor their own intake, domesticated animals had to rely on their owners or carers, who were advised to ensure that sufficient appropriate foodstuffs were available throughout the year.

The fifth non-natural of evacuation also applied equally to humans and animals. It was based on periodical treatments meant to rid the body of excessive humours. Purging was regularly carried out as a preventative measure, as well as to cure diseases. The modern definition of purging refers to the emptying of the bowels. However, in the seventeenth century this was only one of many different methods used to remove unwanted materials from the body. Vomiting, 'neesing' [sneezing] and 'gargarismes' [gargles] were three popular methods for clearing the upper body. The best known, and most commonly employed type of purging was probably phlebotomy, that is, cutting a vein. Bleeding was thought to be especially useful for removing alien matter or excessive chemical principles from the body.

Although there were strict guidelines about the proper timing for bleeding humans, this was not always the case for animals. William Dade recommended making an incision on the neck of horses, and drawing blood on the first day of April to make them stay healthy 'the whole year'. Phlebotomy also needed to be done cautiously as 'the letting of blood is very dangerous, and openeth the way to many grievous infirmities' if not properly administered.[22]

The final non-natural of the 'passions', however, also included a range of emotions from sexual passion and anger through to jealousy. This is another, albeit small example, in diverging advice for humans and animals. In moderation, and at the correct times of the year, human sexual intercourse was an important form of preventative medicine. Most animals, however, were not allowed that liberty. While there was little that could be done about animals that were emotionally out of sorts, it was possible to meddle with their sexual activities, the most dramatic and permanent way of control being castration.

When preventative methods eventually failed, there were a number of different treatments and remedies that could be administered to sick humans and animals. Certain humoral imbalances required the ingestion and retention of organic or inorganic substances. They were thought to be particularly effective when an illness was caused by a humour being 'any less ... than it ought to be'. Such imbalances could be rectified, it was thought, through a mixture of the proper diet, supplemented by medicinal potions. Alternatively, preparations could be used topically for soothing skin, or other external complaints. One recipe for healing broken bones called for comfrey and wild daisies to be boiled in turpentine and yellow wax. The mixture would be made into a 'plaister', which was laid on the 'grief' and supported by 'splinters' to keep it stiff.[23]

Other remedies involved the 'taking away' of offending substances from the body. Galen thought that animals intrinsically understood what herbal remedies to use when they were ill.[24] This is not surprising, as many dogs, for example, can be seen to purge themselves through eating grass and vomiting. Other animals, however, were forced to rely on medicinal remedies or potions. For example, a potion of groundsell, fenugreek, turmeric, aniseed, cumin, garlic, pepper and licorice could be administered to 'open the pipes' of sick horses. A mixture of chopped savine, honey and butter could be used to expel worms in horses.[25]

Although it was believed that humoral imbalances were the cause of illness in both man and beast, recipes for treating disorders showed many differences. For example, both animals and humans could suffer from 'surfeits' of food or drink. An appropriate remedy for a surfeit in horses was based on 'strong beer' mixed with 'wormwood, celedine and herbraces'. Humans, however, were advised to take essence of poppies.[26]

Both people and animals were prone to being bitten or stung by 'a venomous thing'. The former were advised to make a salve of rue and bacon to apply to the wound. According to the sixteenth-century herbalist John Gerard, the use of rue for these purposes dated back to ancient Greece. The remedy for animals with such an affliction was rather more unpleasant sounding, and certainly more odorous. For oxen 'bitten or venomed' a salve was to be made of a rotten egg, soot and bay-salt.[27]

There were a number of herbs that were used differently for human or animal remedies. Arsmart, or water pepper, was one such plant. This herb was under the dominion of Mars, and was said to have cooling and drying properties. Humans were advised to boil three ounces of leek juice and of arsmart to make a potion to soothe 'fluxes of the belly'. Arsmart was also considered useful for patients suffering from 'heat, stoppage or scalding of urine'. For animals, however, the herb was recommended as a restorative for tired horses. Readers of Swan's almanac were advised to rub tired horses with arsmart, and then to lay 'a good handful or two' of the herb under the saddle.[28]

Savin was another herbal remedy that was used differently for human or animal afflictions. It was first mentioned in an almanac of 1684 for killing worms in oxen, cows or calves. It was later suggested that this was good for curing horses and sheep, in addition to cows. The alamanac author Swallow disagreed, writing that it was mixed with honey and butter, and fed only to horses. According to the herbalist Culpeper, this was an easily accessible plant that killed worms in humans, as well. His *Complete Herbal* recommended spreading it on a piece of leather and applying it to the navel. It was apparently also a popular treatment for inducing abortions.[29]

Other afflictions suffered by man and beast required more drastic measures. The most benign sounding remedy for warts on animals was to apply 'black water that stands in the root of an hollow Elm-tree'. Another suggestion for dealing with 'this Disease ... most incident of young Beasts' advised tying eight or ten horsehairs tightly around the wart and leaving them until it fell off. Alternatively, it could be seared off with a hot iron, or it could be eaten away with mercury. Humans, however, were advised to burn away warts with chemical mixtures containing verdigris or acetate of copper. If the reader preferred something milder, the alternative was to rub the juice of celandine into the skin. This remedy is still recommended by herbalists for removing warts and for curing eczema in the present day.[30]

Almanacs also discuss illnesses that were peculiar to animals. 'Murrain' was a frightening animal disease which often proved fatal. It was also known as 'cattle plague' and was later called 'rinderpest'. Murrain was known to be highly contagious, and some towns were said to have the 'common charitable custom' of cutting off the heads of cattle who had died from it, in order to:

> Put it upon a long Pole, and set it on a Hedge fast bound to a stake by the high-way side, that every man that rideth or goeth by that way, may see and know by that sign, that there is sickness of Cattel in that Township.[31]

Clearly, the remedy offered by several seventeenth-century almanac writers based on hens' dung soaked in old, human urine was not the answer. In fact, those who administered the potion were likely to become ill themselves, not with rinderpest but the related, human form of the disease of measles.[32]

## Healers

Not only was human and animal medicine very similar, there were also many similarities in the types of healers who specialized in human and animal medicine.[33] I would argue that there were two, parallel types of medical marketplaces, one for humans and the other for animals. The practitioners at the top of the 'human' market hierarchy were the traditional 'tripartite' division of physicians, surgeons and apothecaries. They were joined by various types of quacks, midwives, clergymen, drug peddlers, magical healers, herbalists, astrologers, dealers in drugs and/or medical appliances and a host of other healers.[34] Most primary care, however, is thought to have been administered within a domestic setting, generally by housewives.

A similar hierarchy can be seen in the animal medical marketplace, beginning with farriers who specialized in horse medicine. Theoretically, the only men who were qualified to use the title were members of the Company of Farriers which began as a guild in 1356. In the seventeenth century, there were many similarities between the Company of Farriers and the College of Physicians. Both were elite institutions with relatively few members. They both also held a monopoly on their respective trades in London and within a seven-mile radius.[35]

As one contemporary author pointed out, however, treating horses was more difficult than treating humans who voice their 'complaints and relations'. Farriers had to deal with 'dumb creatures', and needed to 'hammer out' information by themselves. Their main duty, however, was the same as physicians, which 'consisteth, in two principall points, First to preserve health. Secondly, to cure the sick'. The role of a farrier was to help prevent or cure 'all Diseases, Griefs and Sorrances incident of Horses, with their Symptoms and Causes'.[36] The two main ways in which farriers carried out their duties was either through phlebotomy or with herbal preparations for internal and external use.

Although members of the Company of Farriers were at the top of the hierarchy of animal practitioners, they were not the only men to treat horses. The second major category included 'horse-doctors' that often used the title of farrier, horseleech or leech. In human medicine, the closest parallel was probably those healers referred to as mountebanks or quacks who had no 'professional' qualifications for practising medicine. Such self-styled animal healers, such as leeches, far outnumbered the membership of the Company of Farriers, and were the dominant healers in the countryside. The level of stigma attached to these 'unqualified' practitioners was very probably not as great as it would be today, particularly since there were vast numbers of horses in both town and country who would have required medical assistance. As David Harley has pointed out, consumers were seeking health, rather than a specific type of medical service.[37] In many cases, the expertise of the farrier or horse-leech was likely to have been a more important consideration than any 'legal' qualifications that they might have had.

Popular medical books suggest that the treatments administered by horseleeches were the same as those offered by farriers. The continuing existence of such men suggests that many were perceived to offer effective treatments. As one eighteenth-century writer aptly noted: 'Experience is the only probable means of success in any individual.' The same phenomenon was visible in human health care, whereby 'professional' practitioners offered the same types of services as 'popular' healers.[38]

It seems likely that the majority of treatments were actually administered by laymen at home, both for themselves and their animals. As one contemporary author pointed out, it was important to know how to 'apply [medicines] to himselfe, whereas neither Physician nor Apothecarie can bee had'.[39] What we now call 'primary health care' (for humans) was the responsibility of the female head of the family. Knowledge of physick was seen as one of the housewife's 'principal virtues'. Since women were often responsible for the care of small animals, they might also tend to their illnesses. There is little evidence to suggest that this would have been the case for larger animals, such as horses or oxen, who probably would have been treated by male members of the household.[40]

## Conclusion

This chapter has attempted to show that the concept of 'one medicine' does apply to early modern English society. Charles Schwabe has speculated that this was because human and veterinary medicine had almost identical 'scientific underpinnings'.[41] Since concepts of health and sickness are integral to a people's sense of social reality, the way in which a society treats sick animals can only be understood in relation to the contemporary social and cultural factors.

As Keith Thomas has rightly observed, these beasts were in many ways 'subsidiary members of the human community, bound by mutual self-interest to their owners, who were dependent on their fertility and wellbeing'.[42] As such, their owners and caretakers were obliged to do everything in their power to maintain and promote their health. It therefore made sense to mould and adapt the methods they would use themselves in order to care for their animals. As a result, there were many parallels between human and animal health care. Both depended heavily on the concept of preventative medicine with an emphasis on a daily regime based on the non-naturals. Similarities can also be seen in both the structures of the medical marketplaces for humans and for animals and the types of treatments and other available options. This is one area, however, about which almanacs do not offer abundant information, as they merely show the advisory supply side of the markets.

I would argue that the failure to properly acknowledge the importance of veterinary history has had, and continues to have, serious ramifications not only for our understanding of the evolution of veterinary medicine, but also for its increasing interdependence with human medicine. To acknowledge the fact that human and veterinary medicine had almost indentical 'scientific underpinnings' is, in the end, simply not enough.[43] There is also a huge amount that historians can learn about English society and culture from the ways in which current medical knowledge was used to prevent and treat the ills of man and beast in early modern England.

## Notes

1. R. Porter, 'Man, Animals and Medicine at the Time of the Founding of the Royal Veterinary College', in A. R. Mitchell (ed.), *History of the Healing Professions*, vol. III (London: CAB International, 1993), p. 19.
2. C. W. Schwabe, *Veterinary Medicine and Human Health* (Baltimore: William & Wilkins, 1984), p. 2.
3. E. Cotchin, *The Royal Veterinary College: a Bicentary History* (Buckingham: Barracuda, 1990); R. Dunlop and D. Williams, *Veterinary History* (St Louis: Mosby, 1998); D. Karasszon, *A Concise History of Veterinary Medicine*, trans. E. Farkas (Budapest: Academiai Kiado, 1988); I. Pattison, *The British Veterinary Profession 1791–1948* (London: Allen, 1984); L. Pugh, *From Farriery to Veterinary Medicine 1785–1795* (Cambridge: Cambridge University Press, 1962); F. J. Smithcors, *Evolution of the Veterinary Art: a Narrative Account to 1850* (London: Heffer, 1958); J. Swabe, *The Burden of Beasts: a Historical Sociological Study of Changing Human–Animal Relations and the Rise*

*of the Veterinary Regime* (Amsterdam: University of Amsterdam Press, 1997); L. Wilkinson, *Animals and Disease: an Introduction to the History of Comparative Medicine* (Cambridge: Cambridge University Press, 1992).

4. A. Wear, 'Religious Beliefs and Medicine in Early Modern England', in H. Marland and M. Pelling (eds), *The Task of Healing: Medicine, Religion and Gender in England and the Netherlands 1450–1800* (Rotterdam: Erasmus, 1996), p. 145 and M. Pelling, 'Trade or Profession? Medical Practice in Early Modern England', in *The Common Lot: Sickness, Medical Occupations and the Urban Poor in Early Modern England* (London: Longman, 1998), p. 232.

5. B. Capp, *Astrology and the Popular Press: English Almanacs, 1500–1800* (London: Faber & Faber, 1979), pp. 25–7.

6. P. M. Jones, 'Medicine and Science', in L. Hellinga and J. B. Trapp (eds), *The Cambridge History of the Book in Britain*, vol. III, 1400–1557 (Cambridge: Cambridge University Press, 1999), p. 438; E. Bosanquet, 'English Seventeenth-Century Almanacks', *The Library*, 10 (1933), 368.

7. D. and J. Parker, *A History of Astrology* (London: Deutsch, 1983), p. 152; M. McDonald, 'The Career of Astrological Medicine in England', in O. P. Grell and A. Cunningham (eds), *Religio Medici: Medicine and Religion in Seventeenth-century England* (Aldershot: Scolar, 1996), p. 77; C. Blagden, *The Stationers' Company: a History 1403–1959* (London: Allen and Unwin, 1960), p. 188.

8. A. Chapman, 'Astrological Medicine', in A. Wear (ed.), *Health, Medicine and Mortality in the Sixteenth Century* (Cambridge: Cambridge University Press, 1979), p. 286.

9. R. Allestree, *A New Almanack* (London, 1643), sig. C5r.

10. T. W. Schillhorn van Veen, 'One Medicine: the Dynamic Relationship between Animal and Human Medicine in History and at Present', *Agriculture and Human Values*, 15 (1998), 116.

11. W. Poole, *The Country Farrier* (London, 1652), sig. A1r.

12. W. Lilly, *Anglicus, Peace or No Peace* (London, 1645), p. 27.

13. J. J. Bylebyl, 'Galen on the Non-Natural Causes of Variation in the Pulse', *Bulletin of the History of Medicine*, 45 (1971), 482–5.

14. A. Westwood, *De Variolis & Morbillis: of the Small Pox and Measles* (London, 1656), pp. 15–16; J. Claridge, *The Shepheard's Legacy* (London, 1670), p. 27.

15. M. Holden, *The Womans Almanack* (London, 1688), sig. A3r.

16. W. Lilly, *Merlinus Anglicanus Junior* (London, 1693), sig. B2r.

17. T. Fowle, *Speculum Uranicum* (London, 1680), sig. B4v., J. Booker, *Almanac et Prognosticon* (London, 1642), sig. C2r; Lilly, *Merlinus Anglicanus Junior*, p. 12.

18. L. Mascall, *The Government of Cattle* (London, 1662), p. 6.

19. W. Dade, *A New Almanack and Prognostication* (London, 1684), sig. B3r.

20. Nicholas Culpeper, *Galen's Art of Physick* (London, 1657), p.129; R. Allestree, *The Whole Duty of Man Necessary for all Families* (London, 1690), p. 203.

21. J. O'Hara-May, 'Food or Medicines? A Study in the Relationship Between Foodstuffs and Materia Medica Sixteenth to Nineteenth Centuries', *Transactions of the British Society for the History of Pharmacy*, 1 (1971), 63.

22. W. Dade, *The Country-mans Kalendar* (London, 1700), sig. B4v; E. Pond, *A New Almanack* (Cambridge, 1641), sig. C5v.

23. T. Cocke, *Kitchin-Physick: or, Advice to the Poor, By Way of Dialogue* (London, 1676), p. 9; L. Coelson, *The poor-mans physician and chyrurgion* (London, 1656), p. 53.

24. P. T. Keyser, 'Science and Magic in Galen's Recipes', in A. Debru (ed.), *Galen on Pharmacology* (Leiden: Brill, 1997), p.178.

25. W. Lilly, *Merlini Anglicani Ephemeris* (London, 1659), sig. B2v; J. Partridge, *The Widdowes Treasure Plentifully Furnished with Sundry Secrets: and Approved Secrets in Physicke and Chirurgery* (London, 1631), sig. F5v; Swallow, *A New Almanack* (Cambridge, 1695), sig. B5r.

26. Lilly, *Merlini Anglicani Ephemeris*; W. Dade, *The Country-mans Kalendar* (London, 1698), sig. B3v; W. Salmon, *The London Almanack* (London, 1691), sig. C2v.

27. G. Blunt, *An Almanack* (London, 1657), sig. C4 and J. Gerard, *Gerard's Herbal: John Gerard's Historie of Plants*, ed. M. Woodward (London: Bracken, 1998), p. 266; R. Gardner, *Veterinarium* (1698), sig. A3r.

28. Nicholas Culpeper, *Culpeper's Complete Herbal* (London, 1653; reprint: Ware, 1995), pp. 16–17; R. Johnson, *An Almanack* (London, 1683), sig. B6r; W. Salmon, *The London Almanack* (London, 1696), sig. B6r; and W. Swan, *An Ephemeris or Almanack* (Cambridge, 1657), sig. C6r.

29. W. Dade, *The Country-mans Kalendar* (London, 1684), sig. B3r; E. Pond, *An Almanack* (Cambridge, 1692), sig. C2r; Swallow, *A New Almanack* (Cambridge, 1695), sig. B5r; Nicholas Culpeper, *Herbal*, pp. 234–5; and D. Cressy, *Birth, Marriage & Death: Ritual, Religion, and the Life-cycle in Tudor and Stuart England* (Oxford: Oxford University Press, 1997), p. 49.

30. J. Bucknall, *Calendarium Pastoris: or the Shepherds Almanack* (London, 1675), sig. C2v; W. Dade, *The Countrymans Kalendar* (London, 1692 and 1694), sig. B1v; W. Salmon, *The London Almanack* (London, 1699), sig. B8r; P. Beyerl, *The Master Book of Herbalism* (Custer, WA: Phoenix, 1984), p. 78.

31. R. Trow-Smith, *Livestock Husbandry*, p. 240; and J. B., *The Epitome of the Art of Husbandry* (London, 1670), pp. 93–4.

32. C. Huygelen, 'The Immunisation of Cattle Against Rinderpest in Eighteenth Century Europe', *Medical History*, 41 (1997), 182; W. Dade, *The Country-mans Kalendar* (London, 1683), sig. B3r; W. H. McNeill, *Plagues and Peoples* (Harmondsworth: Penguin, 1976), pp. 54–5.

33. Although there have been some suggestions that the same healers who treated animals also cared for humans, I have found little evidence to support this.

34. A. Wear, 'Medical Practice in Late Seventeenth and Early Eighteenth Century England: Continuity and Union', in R. French and A. Wear (eds), *The Medical Revolution of the Seventeenth Century* (Cambridge: Cambridge University Press, 1989), p. 302.

35. H. J. Cook, *The Decline of the Old Medical Regime in Stuart London* (Ithaca, NY and London: Cornell University Press, 1986), p. 20 and L. B. Prince, *The Farrier and His Craft: the History of the Worshipful Company of Farriers* (London: Allen, 1980), pp. 1–2.

36. R. Elkes, *Approved Medicines of Little Cost* (London, 1652), sig. A2v.; S., *The Gentleman's Compleat Jockey* (London, 1697), sig. A1r.

37. D. Harley, 'The Good Physician and the Godly Doctor: the Exemplary Life of John Tylston of Chester (1663–1699)', *The Seventeenth Century*, 9 (1994), 94.

38. E. Snape, *Snape's Practical Treatise*, p. 1; and D. E. Nagy, *Popular Medicine in Seventeenth-century England* (Bowling Green: Bowling State University Press, 1988), p. 3.

39. R. Trow-Smith, *A History of British Livestock Husbandry to 1700* (London: Routledge & Kegan Paul, 1957), p. 240; O. Wood, *An Alphabetical Book of Physicall Secrets* (London, 1639), sig. A2v.
40. D. Simonton, *A History of European Women's Work 1700 to the Present* (London: Routledge, 1998), p. 20; L. Prince, *Farrier*, p. 227.
41. Schwabe, *Veterinary Medicine*, p. 2.
42. Keith Thomas, *Man and the Natural World: Changing Attitudes in England 1500–1800* (London: Penguin, 1984), p. 98.
43. Schwabe, *Veterinary Medicine*, p. 2.

# 5

# Consulting by Letter in the Eighteenth Century: Mediating the Patient's View?

*Micheline Louis-Courvoisier and Séverine Pilloud*

Medicine today, fortified by a wide variety of highly specific investigative tools, deals largely with physical examination. The human cell has recently been deciphered to its smallest unit, the gene, promising new possibilities for therapy and research. This technical evolution, that first took root in the nineteenth century, has profoundly affected the nature of the patient–doctor relationship. According to some medical historians and sociologists, sick persons' bodies have gradually replaced their narratives; in physicians' eyes, reading pathological signs through the mediation of different instruments has become more relevant than listening to a description of symptoms.[1] The therapeutic interaction tends to confront a speechless patient with a scrutinizing practitioner, especially with the move towards hospitalization. In this particular clinical situation, not only the words but also the world of the patient appears to vanish progressively. Placed in an institution, sick people are cut off from their community.[2]

The situation was quite different with Enlightenment medicine. Diagnosis then was based mainly on an account of the sensations felt by the patient.[3] The history of illness as told to the doctor constituted the main source of information, even if bodies were also occasionally submitted to a more or less extended observation.[4] The fact remains, nonetheless, that medical practice was based largely on verbal testimony, such that the practitioner did not always have to see his patients, relying instead on a written report in order to formulate his opinion: hence the relative importance of doctor–patient correspondence during the eighteenth century.[5]

This chapter is not about the written consultations given by physicians but rather about the epistolary requests addressed to them concerning someone's sickness. Such an archival body of records affords exceptional interest, not so much because it deals with patients but above all because it originates with them. If not necessarily leading to an objective truth, it does, however, give access to a genuine perception that is difficult to grasp through other historical records, which have often been marked by the interests of others concerned with the health system.[6] Much recent work centred on the patient's view has indeed studied the way individuals themselves experience sickness, in their flesh as much as in their mind, seeking different kinds of help to restore their health.[7] This shift in perspective also allows the intervention of other protagonists to be brought out. For, as Roy Porter points out, 'it takes two to make a medical encounter – the sick person as well as the doctor ... it often takes many more than two, because medical events have frequently been complex social rituals involving family and community as well as sufferers and physicians'.[8] The therapeutic relationship appears as a complex interaction, in which different mediators engage. We will deal mainly with the question of mediations within *epistolary* requests for consultation, mediation being understood as an action undertaken by a certain person, which has an influence on the epistolary relationship between the practitioner and the sick person; but the notion of mediation can of course be thought of in a larger sense; it concerns, in a more general way, the handling of sickness in everyday life, within a social group. Focusing in particular on the patient's view or voice, it appears to be occasionally mediated by other actors, as revealed by the correspondence addressed to Samuel Auguste Tissot.[9]

## Dr Tissot, his patients and the mediators

The Swiss physician Tissot was very famous during his lifetime due largely to certain books he published, notably *Onanism: a Treatise Upon the Disorders Produced by Masturbation* (*L'Onanisme*)[10] and *Advice to People in General with Respect to their Health* (*Avis au peuple sur sa santé*).[11] He received a considerable quantity of correspondence, around 1200 letters or *mémoires* each describing the case of a sick person in order to obtain Tissot's advice on diagnosis and treatment.[12] These documents, on which other scholars have already worked, notably Daniel Teysseire[13] and Michael Stolberg,[14] stimulate new reflections on the social history of medicine in the eighteenth century, incorporating as they do individuals' representations and practices in relation to their health and their bodies, expressions of ailments and pain, the patient–practitioner dynamic and therapeutic pluralism. Written between 1765 and 1797, the documents came from all over Europe,[15] but mostly from France, Italy and Switzerland. They were composed by more than 800 different authors, some of whom sent several letters over a particular period of time. They were written by sick people themselves, but also by others, mediators, who will be discussed later. Some patients' files contain several documents (up to thirteen) whereas others consist of only one. About half of the texts are written in letter format, with conventional openings and endings, to which signatures have usually been added. The others are structured as a kind of report or description of the illness, drafted in a less personal tone than in the letter format.

Despite the great heterogeneity of these documents,[16] some common elements can be pointed out: they usually begin by introducing the sick person, often including details related to his or her constitution, temperament and way of life. There follows a list of previous illnesses with information concerning treatment taken and the diagnosis or hypothesis suggested by practitioners previously consulted. Then the author describes the current ailments, the results of different examinations (blood, urine, faeces and pulse) and the different therapies so far attempted. Some explanations or ideas about the sickness and its cure might finally be added before the author formulates his or her hopes and expectations regarding Dr Tissot.

As mentioned above, other people besides the patient were likely to have taken part in writing the documents, and these people could fulfil several types of roles, which illustrate the complexity of mediations in the epistolary relationship. The three major roles for mediators were:

- the mediator-scribe, who acts as a mere secretary, lending his pen and putting words to paper at someone else's dictation;
- the mediator-observer, who has observed the course and the symptoms of another's illness, and reports them, without necessarily writing the request for consultation;
- the mediator-author, who drafts the narrative using his own words, giving his own point of view on the sickness, and referring to the patient in the third person.

Of course, these different levels of mediation could be combined, as in a document written by a Monsieur Dedelay d'Agier concerning a

32-year-old woman, a friend of his wife. The patient has related the story of her sickness to Madame Dedelay d'Agier who, in turn, has dictated the letter to her husband. The married couple functions here as a pair of mediators; he is the mediator-scribe and she the mediator-author.[17] In the letter concerning Monsieur Duvoussin, there are also two kinds of mediation; his father takes on the narration, adding the comments of a mediator-observer, the physician, who having performed two palpations, has discovered tubercles on the liver.[18]

In the discussion that follows, attention will be focused mainly on the mediation of narration, which can be considered as the most important one because it concerns the authorship of the text. The person who speaks in the patient's place and signs the request for consultation may give his or her own interpretations. The crucial question of the subjective position of the mediator-author will not be debated here, nor the kind of alterations inherent to the textual *medium* itself or to the addressee; the account of sickness is indeed undoubtedly determined, at least in part, by the whole process of writing and by the personality or image of Tissot. Rather, the intention here is to try to understand the reason for the patient's silence, and to examine who is telling his or her story, and why.

We have noticed that about 60 per cent of Tissot's patients were not the authors of the letter or the *mémoire* relating their ailments, which raises several questions. For example, two authors, his wife and one of his friends, report the history of Monsieur Chatelain's illness.[19] The document does not explain why the patient cannot tell his story himself. It indicates that his right hand is crippled, but nothing would prevent him from dictating to a secretary. Similarly, when a Monsieur Viton seeks consultation for his wife, one imagines that she is unable to do so herself, or that it is not usual for women to undertake this kind of endeavour. But those suppositions do not fit, because the next letter is actually signed by her: she thanks Dr Tissot for his prescription, and gives additional information about her health.[20] So how has her husband come to write the first letter?[21] One can also ask why Madame D'Hervilly does not draft the document concerning her own sickness (her sister does) although she has previously sent many letters about her daughter's health.[22] If in some cases the reason for epistolary mediation may be understood easily, for example when a parent seeks advice for a young child, in other cases the motives are more difficult to discern. Let us consider some hypotheses.

## Patronage or other interpretations of epistolary mediations

Regarding the patterns of the epistolary relationship between the sick person and the healer, it has been claimed that very few patients approached the doctor directly, except for titled aristocrats or those who had already been in contact with the doctor. Most patients, it is argued, were supposed to do so through intermediaries known to the physician. Laurence Brockliss has qualified this mediation as a kind of *patronage*[23]: in the seventeenth and eighteenth centuries people often had to be introduced and recommended by someone else in order to solicit assistance or information. Examining this first hypothesis on the basis of Tissot's medical correspondence, we have discerned that a mediator does not represent about 40 per cent of his patients when they address their first communication to him. This is a quite significant proportion. We do not have much information about the socio-cultural situation of these patients, but we can affirm that titled aristocrats are not the only ones to act without a mediator, so the hypothesis of patronage is not satisfactory to explain the epistolary mediations.

It has also been asserted that it was common sense to entrust the composition of this written request for consultation to medical men, most often physicians but also surgeons or apothecaries, because they were considered best qualified to do this, even if the social elite to which most of the sick persons consulting by letter belonged, had quite a good command of contemporary medical discourse.[24] The use of lay intermediaries, such as family members, ecclesiastics or acquaintances, was also possible, but it would stand out as an exception, reflecting the patient's isolation, both social and geographical. This second hypothesis concerning the great predominance of *medical mediation* does not fit well either in regard to Tissot's patients. An important number of authors are lay people, and we cannot reduce their intervention to a last resort in the absence of other, better, solutions. The problem is therefore more complex. Other interpretations need to be considered.

First, it is important to make a distinction between the document containing the history of the illness and the letter of introduction; the latter generally does not reveal much about the pathology itself and serves another purpose that remains to be defined more clearly. It should be noted that not all correspondence addressed to Tissot is preceded or accompanied by a letter of introduction (less than 20 per cent) which means that such a procedure was not so common when asking for medical advice. About 65 per cent of these letters of introduction are signed by a mediator, most of the time a family member (about 20 per cent). The authors can also be healers (approximately 15 per cent), ecclesiastics, or friends and acquaintances, among whom some of Tissot's patients, while consulting for their own trouble, take the opportunity to introduce a new patient. Sometimes the intermediary's identity or link with the patient is not revealed.

In any event, the predominant mediation in the letters of introduction appears to be assumed by family members, which invalidates at least partially the hypothesis of patronage. Since close relatives are generally situated at the same social level as the patient, it is highly improbable that they would have acted as influential patrons. Furthermore, more than 35 per cent of the documents are written by the sick person him- or herself, which proves that they cannot be reduced to a means of recommendation.

## Patient's view and writer's view

In fact, the main use of the letter of introduction is to present the patient to, and legitimate his or her approach towards the doctor. It is, however, true that some documents mention the name of other persons, acquaintances or family members, who have been successfully treated by Tissot, or who have come to know him personally. The aim is to position the patient within a recognizable social circle; but we cannot talk about 'vertical patronage' for the sick persons, and the people referred to generally belong to the same network. This is more like a process of identification. The patients are not complete strangers to Tissot once it can be demonstrated that they share common relationships with the doctor.

Now concerning the narrative of the ailments sent to Tissot, we still find about 40 per cent of patient-narrators, who use the first person to designate themselves and to tell their story. But the mediation occurring in the remaining 60 per cent is not of the same nature as in the letters of introduction. The mediator-authors are mainly physicians or surgeons (about 20 per cent) followed by family members (about 10 per cent), friends or acquaintances, and ecclesiastics. There are also some teachers and army officers. The medical mediation is thus predominant when dealing with the story of sickness, which tends to demonstrate that among mediators, medical professionals seem to be preferred to lay authors to describe the symptoms. But one must be careful not to jump to conclusions.

First, it is important to note that medical mediators do not always write to Tissot at another's request. Several of them do so on their own initiative. One example is Dr Millet, who had been trying without success for fifteen months to cure Madame de L'Ecluse.[25] He is positioning himself as the one who needs advice. Dr Metzger is in more or less the same situation with his epileptic patient. But he is also eager to derive personal benefits, which he openly admits. He would like to help the woman, he says, this is the duty of an honest practitioner, but he also wants to find favour with her powerful family and hopes that Tissot's fame will enhance his own reputation. In this case, there is actually a patronage relationship, but it is Dr Metzger who solicits this:

> Beyond an interest in humanity, the first objective of any honest physician, I would like to alleviate the lady's suffering in order to commend myself to a powerful and creditable family. This, Sir, is the reason of my letter. My small reputation strives to profit from the greatness of yours.[26]

Here we have a physician seeking the support of one of his more famous colleagues and trying to get on good terms with the social elite.

Other people, most often family members, also take the initiative in consulting Tissot by letter, for example, the Comtesse de Vougy, who decides to send a letter to the doctor without her husband's knowledge.[27] She is extremely worried about her husband, thinking that he neglects his health because he refuses to follow her advice. She needs Tissot's help to make him more sensible. There is here a disagreement between the married couple, and the physician is implicitly exhorted to resolve the conflict, being himself, in a way, a kind of conjugal mediator. He is sometimes also drawn into familial issues, as when Madame Courtevel de Pezé d'Argouges, who is very upset about the state of her sick daughter, begs him to relieve the young woman of her prejudices against some treatments and to fight her morbid ideas.[28] In this case, too, the letter must remain secret.

These situations are not exceptional. Several authors are critical of the sick person's attitude, notably concerning therapeutic compliance, which demonstrates that if the epistolary requests for consultation offer a certain access to the patient's view, it is not always faithfully translated by the mediators' words. This is not very surprising, for diagnosis and treatments are often at the centre of discussions or debates that imply involvement not only of the practitioner and the patient, but also of different people moving around them. The authors sometimes express personal expectations that do not necessarily converge with those of the sick person, supporting the idea that they are not always invited to fulfil their role, but do so sometimes out of self-interest.

In some cases, however (about 3 per cent of the documents), it is explicitly mentioned that the mediator-author has been asked to draft the text. And if the patient or the family often chose a healer, it must be underlined that they might also decide to appeal to a lay intermediary, for very precise reasons. Monsieur Bon's letter is a good illustration of this kind of mediation; it concerns a young girl, suffering from epilepsy. The author is a relative living in the same house, who has witnessed the fits. He explains that the parents have preferred his writing to that of a professional, because doctors tend to present their own opinion instead of the true facts:[29]

> Bound by friendship and parenthood with her family, living under the same roof, I have witnessed both the circumstances and the patient's crisis. Her parents have therefore asked me to send you as detailed a report as possible about her illness. They prefer my pen to those of physicians whose opinions always show and present facts as they see them.[30]

This kind of criticism of medical narration is not uncommon, and appears in quite a few patients' letters. Monsieur Thomassin, for example, is very clear about this, and he implies also that any lay mediator would fail to describe his ailments properly. He insists on the fact that his sickness is interior, subjective. He is the only one who can talk about it. A doctor would probably give his own interpretation, which is the reason why he decides to narrate the story of his illness:

> My illness is within me, I alone feel it. I believe that I am the only one capable of describing it; that is why I have not used the services of some doctor belonging to the faculty, who, by using the terms of his art, would explain my case less well than my own simple words.[31]

These last instances show that several of Tissot's patients deliberately chose not to ask a medical professional to draft their written request for consultation, preferring to narrate their sickness themselves or to entrust a lay author to draw up the text. So why do lay or professional intermediaries intervene?

## Epistolary mediations: a complex phenomenon

In some cases the answer seems simple, especially as noted before, in the case of children (about 8 per cent of Tissot's patients) who are unlikely to put their ailments into words. One must also take into consideration the patients' state of health. Those who it is supposed suffer from madness, for example, are usually not able to describe their situation themselves, all the more because they may not recognize their insanity.

The mediation is seldom due to patients' illiteracy because most of Tissot's clients belong to the social elite. There are, however, some exceptions, notably one or two servants who are represented by their employers, among them an aristocrat. In these few cases, we can actually say that the mediation functions as a form of patronage. But it should be reiterated that the hypothesis of patronage does not fit in many other cases. Further thought needs to be given to the question of mediations within the epistolary relationship between doctor and patient.

First, it is wrong to take for granted that it is always the sick person's idea to look for the healer's advice or help. As has been shown above, the initiative may come from other people, even at times acting in secret. So the mediation is not necessarily a service requested by the patient to obtain something. It can, on the contrary, be the sign of some kind of disagreement between the patient and the author, the latter being willing to give his or her own point of view or forcing the sick person into an endeavour he or she would not have undertaken without a third party's intervention.

Besides, and this second point is linked closely to the former, it is essential to consider the impact of sickness on the community, especially on the family. The importance of kin intermediaries proves that relatives often feel very concerned by what is happening to their nearest and dearest. Many of them are upset and eager to be active, to do something to ease pain or to relieve their own anguish. Hence they dare to write to the doctor, revealing the different symptoms and sometimes proposing diagnosis, treatments, or even criticizing the way a physician has handled the case up till now.

One more point must be added, to qualify the hypothesis of medical mediation: lay people, including the patients themselves, feel authorized to talk about medical matters and do it relatively easily most of the time. Though it may not be a simple thing to do, they do not seem to agree with the idea that discussing sickness is the doctors' privilege.

Some of them, as mentioned before, even mistrust professional narration, considering it too dogmatic and prejudiced, which might distort the reality of the symptoms. This offers some original clues concerning the physician's status in the eighteenth century, a subject that will not be developed in this chapter.[32] There are indeed some quite interesting aspects of the social history of medicine that are tackled by reflection on the epistolary mediations, as will be presented by way of conclusion.

First, in the field of medical practice, a common assumption is being confirmed: it is true that there are many lay actors who take on the function of care substitutes, intervening as mediators between a doctor and a sick person. For instance, the Marquis de Cély and his wife appear to have sent several letters about patients in their area; one document even says that the man threatened to withdraw his help from a woman if she did not show more compliance towards Dr Tissot's prescription.[33]

Madame Fontanes represents another example of lay medical adviser, functioning as an intermediary between doctor and patients. She consults Tissot for the daughter of a friend of hers, offers a very theoretical explanation of the sickness, conveys the opinion of some healers, evaluates the treatments prescribed and, having read a treatise by Tissot on nervous sicknesses, gives her own opinion about the nature of the symptoms:[34]

Sir, the case in question seems to be relevant to your treatise on epilepsy, but having not found a comparable example ... I strongly recommended writing to you ... and offered my services as a secretary. With your intelligence, you will be able to complete this description, and we can later add any further information you may require.[35]

These two illustrations are enough to show that some mediators not only took on the narration of the sickness, but also contributed by helping the people in their neighbourhood. Tissot had intended his famous book, *Advice to People in General with Respect to their Health*, precisely for just such social intermediaries, who belonged to the educated elite and thus were capable of reading, and who lived in the country, next to peasants or other less privileged people.[36] It was above all aimed at the ministers, and it would be very interesting indeed to analyse more thoroughly the healing function of the churchmen in their parishes. But it was also directed to the aristocrats or any other reasonably well-off people, schoolteachers, army officers or representatives of liberal professions. This treatise was intended to be used in order to provide first aid or counsel in the absence of professional physicians, who were not always within reach in small villages.[37]

Focusing on the epistolary mediations also provides, as mentioned before, some interesting insights regarding the experience of sickness by the social environment of the patient. The anxiety of a parent or a friend is often very perceptible, as in the letters from Madame Decheppe de Morville about her husband. This file shows how his wife observes the course of the disease day and night, giving many details about the quantity and quality of his urine, the colour of his skin, his loss of weight, the swelling of his leg, the nature of his vomiting and so on:

> Although he quite regularly urinates once supper is finished, it is only at about four in the morning and before getting up that he does so abundantly. His urine appears to be of natural colour and not superior to his liquid intake.

Being next to her suffering husband and observing him all the time leads her to elaborate a diagnosis: she is afraid of him being affected by dropsy and notes with fear that his obstructions have not disappeared as they should have. This latest stage in the illness has redoubled her fright and her sorrow:

> I confess that this last accident has increased my fears and my sorrow ... I tremble at the thought that the swelling could be a symptom of the dropsy, and that the obstructions are not dissolved as they seemed to be.

In February 1783, she writes that her anguish and grief increase every day. This is partly related to the distance that separates her from Tissot; he is the only one who has her complete trust and who can ease her worries.

> Please forgive my anxiety, which has increased every day since the various accidents that my husband has recently suffered, consider my trouble, living so far from you, the only person in whom I have confidence; I wish I could hear you at all times, you are the only one who could calm me down.

Monsieur Decheppe de Morville is not improving, and in March 1784, his wife writes that his state affects her so much that she cannot convey it, all the more so because she has not yet received an answer from the physician. She has the impression of being left all alone with her torments: 'Sir, I await your advice and council concerning the condition of M. de Morville. His present state preoccupies me beyond what I can express, and your directions are my only hope.' This collection of documents concerning Monsieur Decheppe de Morville ends abruptly with a report of a post-mortem examination.[38]

## Straightforwardness, shame or secret in social discourse about the body and illness

Generally speaking, patients' files, which sometimes contain several documents written by different people but about the same person, are very interesting in tracing the course of the symptoms and the way such an evolution is lived by the surrounding community. They also bring to light the various and subjective representations, or attitudes, towards health. The Prince of Piémont's file serves as a good illustration: among seven documents, two are signed by the aristocrat himself, the others being composed by his practitioner and his wife, who are both very worried about him. The woman notably explains that she has insisted that the Prince writes himself to Tissot, but for quite a long time he has been unwilling to do so, being very happy with his treatment and not seeing the necessity of another consultation.

> I consistently insisted that he write to you himself, after the period of twenty days, but to no avail since he declared that his regimen agreed with him too much for him to try another. In October, he refused to take any precaution against the cold.[39]

The woman has a different conception of hygiene and prevention to that of her husband. Convinced that he is not able to take responsibility for himself, she is led to take part in the therapeutic endeavour, addressing several letters to Tissot in place of the patient, who is too reluctant to do so regularly. Such a conjugal mediation demonstrates the extent to which illness affects individuals' interactions with each other, being the source of negotiation between the different parties concerned.

The norms or values underlying human relationships are indeed one more aspect enlightened by reflection on the epistolary mediations, offering new perspectives on social and cultural history, around themes like privacy or sense of decency. For instance, when Madame Gounon Laborde writes, at her husband's dictation, a document concerning his ailments, she cannot avoid the question of his sexuality. Going back over his past, he relates, but she writes, that he has always indulged his passions, having loved women desperately, drunk too much, and gambled all his life:

> I have repeatedly suffered from all types of passions, I have loved women furiously. I have gambled all my life and drank much wine and consumed coffee continuously. My nature has always been to take all that comes with excess. For as long as I can remember, I have almost always taken medicine for my nerves which have always felt irritated and which my cruel passions have irritated yet more.[40]

Monsieur Gounon Laborde is disclosing very personal matters through the mediation of his wife, which implies that the couple do not have many secrets from each other. We could draw similar conclusions when reading Monsieur Martin's letter about Monsieur Demeunier's onanism[41] or when following the Marquis d'Albaray's descriptions of the Comtesse de Mouroux's menstruation[42] and complaints related to the genital organs. There is undoubtedly something to say about the various thresholds of sensitivity and the idea of intimacy in the eighteenth century, especially regarding different kinds of pathology. It would be interesting to explore the link one may find, if any, between the nature of the sickness and the profile of the mediator (gender, age, medical or lay intermediary, for instance). At first glance, it does not appear that complaints connected with menses, menopause or pregnancy are especially related by women, but this issue deserves a more precise study. It is true, however, that the theme of onanism is not treated in a banal way. Many men emphasize the difficulty of such a confession. If some of them seem almost relieved to tell all, others choose to remain anonymous, sometimes asking someone else to represent them. Monsieur Ducassé explains that one of his friends has asked him to report this 'horrible crime'.[43] But the amount of detail contained in his letter is rather surprising, and one wonders if the author is not in fact telling his own story, trying to protect his dignity behind a 'virtual mediator'.

This last instance underlines the intricacy of the mediations in the epistolary relationship between the patient and the practitioner, which reflects the complexity of the questions raised by the eruption of illness, regarding not only the sick person him- or herself but also the surrounding people. Such an archival body demonstrates a fact that is perhaps a universal truth: the experience of sickness is not the business of one individual alone, it often implicates third parties. It is even truer concerning the eighteenth century, because of the particular structure of the medical market and institutions. At that time, when hospitalization was quite rare and professional services not so easily accessible, relatives were usually very involved in the therapeutic process. Help and care at home played an essential part, all the more so because people often had to wait before consulting a physician or any other healer. Journeys took quite a long time, as did epistolary exchanges, which had to come and go before being of any use. Hence the crucial importance of the community in the process of coping with sickness.

## Notes

1. M. Fissel, 'The Patient's Narrative and Hospital Medicine', in R. French and A. Wear (eds), *British Medicine in an Age of Reform* (London and New York: Routledge, 1991), pp. 92–109; J. Lachmund, 'Between Scrutiny and Treatment: Physical Diagnosis and the Restructuring of 19th Century Medical Practice', *Sociology of Health and Illness*, 20 (1998), 779–801.
2. N. Jewson, 'The Disappearance of the Sick Man from Medical Cosmology, 1770–1870', *Sociology*, 10 (1976), 225–44.
3. See notably V. Barras and Ph. Rieder, 'Ecrire sa maladie au Siècle des Lumières', in V. Barras and M. Louis-Courvoisier (eds), *La Médecine des Lumières: Tout autour de Samuel-Auguste Tissot* (Geneva: Georg Bibliothèque d'Histoire des Sciences, 2001), pp. 201–22; S. Pilloud, 'Mettre les maux en mots; médiations dans la consultation épistolaire au XVIIIᵉ siècle: les malades du Dr Tissot (1728–1797)', *Bulletin canadien d'histoire de la médecine*, 16 (1999), 215–45; S. Pilloud and M. Louis-Courvoisier, 'The Intimate Experience of the Body in the 18th Century: Between Interiority and Exteriority', *Medical History*, 47 (2003), 451–72.
4. D. Porter and R. Porter, *Patient's Progress: Doctors and Doctoring in Eighteenth-century England* (Cambridge: Polity Press, 1989), p. 77. R. Porter and W. F. Bynum (eds), *Medicine and the Five Senses* (Cambridge: Cambridge University Press, 1993). However, it should be specified that touching the body was not as marginalized an activity as commonly claimed. Quite a number of documents (about 15 per cent) mention palpation or examinations, including several vaginal and rectal ones; how and by whom the body of the patient has been palpated is specified and the result of this procedure seems to be an important element to relate to Tissot. Cf. O. Keel, 'Percussion et diagnostic physique en Grande Bretagne au 18ᵉ siècle: l'exemple d'Alexander Monro secundus' (Bologne: Actes du XXXI

Congresso Internazionale di Storia della Medicina, 1988), pp. 869–75, and 'L'essor de l'anatomie pathologique et de la clinique en Europe de 1750 à 1800: nouveau bilan', in Barras and Louis-Courvoisier, *La médecine des Lumières*, pp. 69–91.

5. Cf. D. Porter and R. Porter, *Patient's Progress*, pp. 76–81; I. Loudon, *Medical Care and the General Practitioner 1750–1850* (Oxford: Clarendon, 1986); G. B. Risse, 'Cullen as Clinician Organisation and Strategies of an Eighteenth Century Medical Practice' in A. Doig, J. P. S. Ferguson, I. A. Milne and R. Passmore (eds), *William Cullen and the Eighteenth Century Medical World* (Edinburgh: Edinburgh University Press, 1993); L. Brockliss, 'The Medical Practice of Etienne-François Geoffroy', in A. La Berge and M. Feingold (eds), *French Medical Culture in the Nineteenth Century* (Amsterdam/Atlanta: Rodopi, 1994), pp. 79–117; L. Brockliss, 'Les membres du corps médical comme correspondants: les médecins francophones et la République des Lettres du 18ᵉ siècle', in Barras and Louis-Courvoisier, *La médecine des Lumières*, pp. 151–69 ; W. Wild, 'Doctor–Patient Correspondence in 18th Century Britain: a Change in Rhetoric and Relationship', in T. Erwin and O. Mostefai (eds), *Studies in Eighteenth-Century Culture* (Baltimore and London: Johns Hopkins University Press, 2000), pp. 47–64; J. Lane, '"The Doctor scolds me": the Diaries and Correspondence of Patients in Eighteenth Century England', in R. Porter (ed.), *Patients and Practitioners: Lay Perceptions of Medicine in Pre-industrial Society* (Cambridge: Cambridge University Press, 1985), pp. 205–48; E. Foster, 'From the Patient's Point of View: Illness and Health in the Letters of Liselotte von der Pfalz (1652–1722)', *Bulletin of the History of Medicine*, 60 (1986), 297–320.

6. E. Wolff, 'Perspectives on Patients' History: Methodological Considerations on the Example of Recent German-Speaking Literature', *Bulletin canadien d'histoire de la médecine*, 15 (1998), 207–28.

7. R. Porter (ed.), *Patients and Practitioners*; E. Wolff, 'Perspectives on Patients' History'. Other kinds of archives can provide this perspective on the patient's point of view, such as family archives, diaries or autobiographies.

8. R. Porter, 'The Patient's View. Doing Medical History from Below', *Theory and Society*, 14 (1985), p. 175.

9. Tissot's medical correspondence, conserved in the public library of the University of Lausanne, contains about 1250 requests for consultation. (Apart from a few exceptions, we do not have his complete reply, but he made notes on quite a number of documents, so we sometimes find his diagnosis or his prescription concerning a particular case.) Realizing the major historical interest of this collection, we decided to try to make it more easily accessible to scholars. With funding from the Swiss National Fund for Scientific Research, we have developed a database, integrating as much information as possible. With around fifty headings for each document, the challenge has been to find a way to simultaneously classify several types of data (such as gender, age and type of sickness) to pave the way for a quantitative approach, while at the same time preserving the particularities of each document. A CD-rom of the database will soon be available. This project (FNRS no. 11-56771.99) has been conducted under the direction of Professor Vincent Barras (Institut Universitaire d'Histoire de la Médecine et de la Santé Publique, Lausanne, Switzerland). Further

publications are planned; the issues addressed deal notably with body experience, therapeutic relationship, lay representation of health and sickness, self-treatment and healing practices at large.

10. S. A. Tissot, *L'Onanisme* (Lausanne: Grasset, 1760, 1st edn.); in English: *Onanism: a Treatise Upon the Disorders Produced by Masturbation* (London, 1766). This book seems to have reached a large audience; about fifty-five documents are related to the question of masturbation.

11. Tissot's most famous books include *L'inoculation justifiée, avec un essai sur la muë de la voix* (Lausanne: Bousquet, 1754, 1st edn); *Avis au peuple sur sa santé* (Lausanne: Grasset, 1761, 1st edn); this text was a bestseller, between 1761 and 1792 it was republished eighteen times and was translated into different languages. Other publications by Tissot include: *Traité des nerfs et de leurs maladies* (Lausanne: Chapuis, 1778–80, four volumes); *De la santé des gens de lettres* (Lausanne: Grasset, 1768, 1st edn); *Essai sur les maladies des gens du monde* (Lausanne: Grasset, 1768 , 1st edn).

12. Pilloud, 'Mettre les maux en mots'.

13. Notably D. Teysseire, *Obèse et impuissant: le dossier médical d'Elie de Beaumont, 1765–1776* (Grenoble: Million, 1995); 'Le réseau européen des consultants d'un médecin des Lumières: Tissot, (1728–1797)', in *Diffusion du savoir et affrontement des idées, 1600–1770* (Montbrison: Association du Centre Culturel de la Ville de Montbrison, 1993), pp. 253–67; 'Mort du roi et troubles féminins: le premier valet de chambre de Louis XV consulte Tissot pour sa jeune femme (mai 1776)', in H. Holzhey and U. Boschung (eds), *Santé et maladie au XVIIIᵉ siècle* (Amsterdam/Atlanta: Rodopi, 1995), pp. 49–56. In his book *Obèse et impuissant*, Teysseire has published the entire file concerning Elie de Beaumont, thirteen documents in all.

14. Notably M. Stolberg, '"Mein äskulapisches Orakel!", Patientenbriefe als Quelle einer Kulturgeschichte der Körper- und Krankheitserfahrung im 18. Jahrhundert', *Österreichische Zeitschrift für Geschichtswissenschaften*, 7 (1996), 385–404; 'La négociation du régime et de la thérapie dans la pratique médicale du XVIIIᵉ siècle Jahrhundert', in O. Faure (dir.), *Les thérapeutiques: savoirs et usages* (Oullins: Fondation Marcel Merieux, 1999), pp. 357–68; 'The Monthly Malady: a History of Premenstrual Suffering', *Medical History*, 44 (2000), 301–22; 'Un Unmanly Vice: Self-Pollution, Anxiety and the Body in the Eighteenth Century', *Social History of Medicine*, 13 (2000), 1–21.

15. The Netherlands, Austria, Germany, Great Britain, Ireland, Scotland, Denmark, Greece, Portugal, Spain, Luxembourg, Russia, Croatia.

16. The number of pages varies from one to thirty-five. Some documents are very precise, with lots of detail, while others are much more sober, and seem to be written as if in an emergency.

17. Lausanne, Bibliothèque Cantonale Universitaire, manuscript department (from now: BCU), IS/3784/II/144.04.04.19, undated.

18. Lausanne, BCU, IS/3784/II/144.01.08.04, 13 January 1772.

19. Lausanne, BCU, IS/3784/II/144.01.05.01, 1770.

20. Lausanne, BCU, IS 3784/II/144.02.06.35 and 36, 25 March and 6 September 1775.

21. Furthermore, we have noticed that quite a number of the authors are women, writing for themselves or for their relatives and acquaintances, which suggests that it was not exceptional for them to do so.

22. Lausanne, BCU, IS/3784/II/144.02.02.16–24, 1770–4.
23. The term patronage, as used by Brockliss, refers to two types of situations: first, when a lay person seeks to obtain a consultation by letter with the help of an intermediary whose name is familiar to the physician; second, when a physician writes to one of his famous colleagues in order to get medical advice, which was considered as a 'mark of deference, a way of acknowledging the superiority of a handful of medical stars. In an age of patronage, it was the means whereby a junior doctor could place his foot on the first rung of the medical ladder.' Brockliss, 'The Medical Practice of Etienne-François Geoffroy', in *French Medical Culture*, pp. 81, 87. In this chapter, we use the notion of patronage mainly in the first sense, that is when a patient is represented by a third party when addressing his/her request to Tissot. But it should be noted that patronage in the eighteenth century has been analysed from other perspectives, particularly in cases in which we find physicians themselves being patronized by aristocratic patients, dependent on their fees and favours. See notably B. Moran (ed.), *Patronage and Institutions: Science, Technology and Medicine at the European Court 1500–1750* (Rochester: Boydell Press, 1991); N. Jewson, 'Medical Knowledge and the Patronage System in 18th Century England', *Sociology*, 8 (1974), 369–85.
24. Eighteenth-century medicine was largely based on humoralism, a concept inherited from antiquity. Its theoretical references were not too specialized or technical, and were thus easily assimilated by the educated people of the upper social classes. See notably R. Porter, 'Laymen, Doctors and Medical Knowledge in the Eighteenth Century: the Evidence of the *Gentleman's Magazine*', in *Patients and Practitioners*, pp. 283–412; A. Wear, 'Interfaces: Perceptions of Health and Illness in Early Modern England', in R. Porter and A. Wear (eds), *Problems and Methods in the History of Medicine* (London, New York: Croom Helm, 1987), pp. 230–55; M. Ramsey, 'The Popularisation of Medicine in France, 1650–1900', in R. Porter (ed.), *The Popularisation of Medicine 1650–1850* (London and New York: Routledge, 1992).
25. Lausanne, BCU, IS/3784/II/144.04.05.18, undated.
26. Lausanne, BCU, IS/3784/II/144.02.05.16, 23 November 1774.
27. Lausanne, BCU, IS/3784/II/144.03.06.35, 3 April 1785.
28. Lausanne, BCU, IS/3784/II/144.05.04, 18 and 21 February 1792.
29. The *mémoire* composed by Monsieur Bon is extremely clear and precise. It is ten pages long and structured in several different parts that expose the origin of the sickness, the symptoms, the treatments prescribed with their results, and so on. Towards the end of his report, the author even inserts some explanatory hypotheses. Such a document would refute the argument denigrating the scientific value of a lay description of illness compared with a professional one.
30. Lausanne, BCU, IS/3784/II/144.05.02.35–6, April 1790.
31. Lausanne, BCU, IS/3784/II/144.02.08.13, 13 March 1775.
32. On this subject, see M. Louis-Courvoisier, 'Le malade et son médecin: le cadre de la relation thérapeutique dans la deuxième moitié du XVIIIe siècle', *Bulletin canadien d'histoire de la médecine*, 18 (2001), 277–96.
33. Lausanne, BCU, IS/3784/II/146.01.02.05, 8 August 1785.
34. S. A. Tissot, *Traité des nerfs et de leurs maladies*.

35. Lausanne, BCU, IS/3784/II/149.01.03.09, 29 December 1774. Madame Fontanes is referring to the *Traité de l'épilepsie,* third volume of the *Traité des nerfs.*

36. S. A. Tissot, *Avis au peuple sur sa santé,* D.Teysseire and C. Verry-Jolivet (eds) (Paris: Quai Voltaire, 1993), pp. 50–2.

37. The appearance of this book, in 1761, undoubtedly increased Tissot's clients, especially those who asked for his services by correspondence. But other publications are referred to in the epistolary requests for consultation, notably *Onanism.* Quite a number of documents are related to the question of masturbation whereas a good proportion deal with complaints described in Tissot's other works – epilepsy, nervous diseases, ailments particular to the *gens du monde* and the *gens de lettres.* See S. A. Tissot, *Traité des nerfs et de leurs maladie; De la santé des gens de lettres; Essai sur les maladies des gens du monde.*

38. Lausanne, BCU, IS/3784/II/144.03.02.06–15, 1783–4.

39. Lausanne, BCU, IS/3784/II/144.05.04.01–07, 1790–2.

40. Lausanne, BCU, IS/3784/II/144.02.02.08, undated.

41. Lausanne, BCU, IS/3784/II/144.04.06.03, undated.

42. Menstruation appears as a constant preoccupation in the documents regarding women. On medical and lay perception of menstruation in the eighteenth century, see Alexandra Lord, 'The Great *Arcana* of the Deity: Menstruation and Menstrual Disorders in Eighteenth-century British Medical Thought', *Bulletin of the History of Medicine,* 73 (1999), 38–63.

43. Lausanne, BCU, IS/3784/II/144.02.04.29, 27 July 1774.

# 6
# Medical Popularization and the Patient in the Eighteenth Century

*Michael Stolberg*

Issues of 'popularization' and 'public understanding of science' have attracted considerable interest in recent historical and sociological writing.[1] Work on the history of medical popularization, in particular, has so far focused almost exclusively on 'popular' medical texts and their (often academic) authors, however. Extant studies tend to present bio-bibliographical data and editorial information, and some of them proceed to embark on a provocative analysis of rhetorical strategies, implicit agendas and ideological backgrounds.[2] This kind of work can throw a welcome light on the authors' intentions and the readership they anticipated. It can tell us very little, however, about the actual impact of such 'popularizing' texts on the general public, let alone on readers of different occupation, class, education and gender.[3] These texts do not reveal how they were read and used, in what way they influenced the medical ideas, the illness experience and the coping strategies of their prospective readers and how influential they were compared to other sources of medical knowledge.

In this chapter, I therefore want to propose a wider approach. I suggest we understand medical 'popularization' not primarily as an intentional act, as something which popularizers 'do'. Rather, I will take 'popularization' to refer to a process which concerns, first of all, knowledge and practices, not authors and texts. From this perspective, the historical analysis of medical popularization calls for the study of the various pathways along which the ideas, the theories and practices of an esoteric medical elite – academic physicians in the Western context – were mediated to and disseminated among a wider lay public and of the ways in which this often heterogeneous public reacted to these ideas and practices, adopting, transforming or rejecting them, as the case may be.

My focus will be on eighteenth-century Europe, when 'popular' works on the prevention and cure of disease came to rank among the bestselling literary genres, making this period a particularly attractive choice for the study of medical popularization. Books like Samuel Auguste Tissot's *Avis au peuple* (1761) or William Buchan's *Domestic Medicine* (1765) went through dozens of editions and translations.[4] And there were hundreds of similar, only somewhat less successful or conspicuous works providing dietetic and/or therapeutic advice for the layman, such as the *Traité des maladies les plus fréquentes et des remèdes propres à les guérir* by Jean-Adrien Helvétius,[5] the *Sicherer und geschwinder Artzt* by the Swiss Theodor Zwinger,[6] the *Leib-Medicus der Studenten* by the German Heinrich C. Abel,[7] or the *Dictionnaire portatif de santé* by Ch.-Augustin Vandermonde.[8] My principal source will be contemporary 'patient letters', or, more precisely, many hundreds of letter consultations, which the sick or their friends or relatives[9] addressed to famous physicians of their time.[10] Clearly, patient letters reflect above all the views and experiences of the literate and affluent classes. Consultation by letter was expensive, and it implied a certain trust in the superior expertise of learned physicians. For the specific purpose of this chapter, however, this is no serious drawback. After all, the literate middle and upper classes were precisely those classes to which the medical advice books were primarily addressed. If these books had any impact at all, it is here that we should be able to trace it.

## Medical lay knowledge

Already a cursory glance at eighteenth-century patient letters amply confirms findings from other sources which show that many educated middle- and upper-class men and women possessed fairly sophisticated and detailed medical knowledge. The extent of that knowledge varied, but generally speaking the cognitive distance between medical expert knowledge and lay knowledge was much smaller than today.[11] Countless patient letters show how closely acquainted the sick and their relatives were with the major physiological and pathological concepts of their time and how well they knew how to apply them to the individual case. Presumably, the acquisition and application of medical knowledge was facilitated by the relatively simple basic explanatory framework of contemporary medicine and its closeness to the everyday experience of the body, which had, in turn, been shaped for centuries by this framework. New medical theories like iatrochemistry, mechanism, Stahlian animism or vitalism put forward distinctly different conceptions of the human body and its relation to mind and soul. But in its more practical aspects, in its understanding of what happened in the sick body and in its therapeutic approaches academic medicine was still largely shaped by humoral pathology. Certainly, the lay understanding of the body and its diseases still moved largely within the traditional framework of humours, spirits and vapours.

It is a widespread misconception that seventeenth- and eighteenth-century humoral pathology was based primarily on the idea of a balance between the natural humours or the corresponding elementary qualities (hot, cold, wet and dry). But by the seventeenth century the principal key to the understanding and treatment of most diseases, among lay persons and physicians alike, was the identification of an impure, morbid matter or humour with more or less specific pathological effects.[12] Such matter might be taken in with food or air, as in the case of contagia and miasms, it might result from insufficient 'concoction' of raw food, or it might stem from the corruption of humours within the body, due to an obstructed humoral flow and, above all, a diminished or totally suppressed evacuation of waste matter. When the morbid humour or harmful vapours rising from it spread throughout the whole body, fevers and other general diseases followed. 'Fluxes' and deposits of morbid matter in individual parts of the body caused local affections like catarrh, gout or rheumatism. With time, they might eventually harden into 'scirrhi' or cancer, that 'most fearful of all diseases', as some patients called it. Thick, viscous humours also might block the vessels and passages, leading to local obstructions and eventually to humoral corruption. Generally, however, experience taught the patients to perceive the morbid matter as extremely mobile, sometimes causing totally different symptoms depending on its site. Accordingly, many patients reported how one symptom, say nasal discharge, disappeared just to be substituted by another – say abdominal pain. As they saw it, the morbid matter in question was still the same. It had just 'thrown' itself on to another part of the body. The ultimate aim in most cures was therefore to mobilize and excrete the morbid matter. Evacuative methods ranging from blood-letting, purgatives and fontanelles to emmenagogues, mineral water and warm baths (to open the pores) continued to provide the mainstay of medical therapy. Modern readers sometimes wonder why the inefficacy of standard therapies like blood-letting was not obvious to contemporaries. Their validity was seen to be constantly reconfirmed by personal experience, however. Countless patients reported, for example, a marked improvement of their health right after a copious evacuation of sweat or faeces – or a drastic deterioration when a habitual evacuation, say, from haemorrhoids or with the menses was 'suppressed'.

Many patient letters also give evidence of detailed knowledge of contemporary physicians' ideas about the causation and prevention of disease. The patients might not heed the rules of healthy living but they were clearly aware that physicians stressed the importance of good and healthy food, of the dangers of culinary and alcoholic excesses, of bad air and abrupt changes of temperature (which might suppress perspiration), of the potentially fatal effects of violent emotions, in short of the traditional canon of the six 'non-naturals'. Accordingly, many patients pointed out a specific event or dietetic error, often many years back, which, in hindsight, had sparked the disease process – for example, a horse ride in the morning dew, undue pressure when they were lifted out of a boat or the terror or overwhelming joy at hearing unexpected news.

On the basis of their often quite sophisticated diagnostic, prophylactic and therapeutic knowledge, the patients and those around them frequently speculated upon the nature and causes of the disease in question. They suggested specific therapies and regimen and sometimes basically just wanted the physician's confirmation that their choice was correct. And they argued with the physician about the validity of his approach, when they did not like it.

## Medical advice books

As the source of their medical knowledge, dozens of patients and relatives explicitly mentioned or even quoted medical advice books and similar popularizing works. In most cases – this goes particularly for the letters to Tissot, one of the most famous representatives of this genre[13] – they referred to works by the physician whom they addressed. Usually these works were quoted on a specific point only, as a source of specific information on a particular symptom, disease or therapy, however, rather than for the general physiological or pathological information they contained. Sometimes diagnosis was the principal issue. A friend had given him Tissot's *Avis*, a French colonel wrote, and he had benefited greatly from it: 'I read, I examined, I checked all the diseases of the chest, I thought I had found my disease among the obstructions, and took this as the starting point to be my own physician.' He chose his therapy accordingly, and mineral water, whey and cold baths helped him recover and turned him back into the vigorous man he had once been.[14] Other patients expressed their astonishment to find their own symptoms described so clearly. 'Your works depicted the state in which I find myself', a German patient wrote.[15] Having read Tissot's explanation of the fainting spells in nervous diseases, a priest from St Malo was convinced that his own fainting spells came 'from the nerves'.[16] Sometimes patients described medical advice books even as the ultimate source of a truth long sought, though factual account and eulogy are not always easy to distinguish in such cases. A 29-year-old man, for example, suffered from severe anxiety attacks and described the 'rather peculiar state I have been in for many years and about which I found elucidation only in your work on the health of literary person'.[17] Along similar lines self-confessed victims of post-masturbation disease declared emphatically how Tissot's *Onanisme* had finally opened their eyes.

Other letters mentioned medical advice books primarily as a source of therapeutic advice. A French priest took alum and, on his physician's counsel, quinine, 'as the famous Mr Thissot (sic) precribes it for weak stomachs'.[18] Another treated his acute fever on the basis of Tissot's *Avis* and now turned directly to Tissot because he could not find any suggestion in his book on the diseases of the 'people of fashion', how to treat the 'pulmonary tubercles' he thought he might be suffering from.[19] A 38-year-old lady successfully cured the severe toothache which she developed when her periods stopped for two months by following the advice in Tissot's *Avis* to let blood on the foot in these cases; her periods returned as well.[20] The Marquise de Louvois had not only cured herself of dysentery and fever on the basis of Tissot's *Avis* but also saved her 7-month-old son when the physicians had already despaired of his life.[21] Tissot in particular also sometimes reached the declared aim of many medical advice books, namely to enable charitable members of the upper classes to provide basic medical care to the less affluent, where there was no physician at hand, especially in the countryside. The Countess of Vougy reported gratefully how she gave first aid 'to the unfortunate sick of my lands, enlightened by the guidance which I found in your advice to the people', and how happy it made her, when the good folks came to her and told her she had saved their lives.[22] Another lady, of noble blood but in precarious economic conditions, asked Tissot for a free copy to support her efforts to help the poor.[23] Some patients also adapted their diet or lifestyle following the counsel laid out in Tissot's works. They took to riding or other forms of physical exercise, slept in cool, well-aired rooms or ate fruit and milk products and so forth.[24]

Since patient letters were usually motivated by a specific case of illness it comes as no surprise that the patients and their relatives referred to medical advice books primarily as a source of practical, immediately applicable, diagnostic and therapeutic information. Other types of sources show that some contemporaries sought and acquired fairly extensive medical knowledge from medical advice books, on a wide range of issues and without direct reference to its immediate practical relevance. The former court-lady and later novelist Mme de Graffigny, for example, was well-read in medical matters, as her letters show.[25] Mme de Maraise even boasted that all she basically needed in order to join the physicians' ranks, was a big wig, a black coat and 'certain great words'.[26] Still, one wonders how often medical advice books, especially those of the 'be-your-own-physician' rather than the 'long-and-healthy-life' type, were actually read more or less from cover to cover, rather than serving only as a potential reference book in times of need. Highly selective reading of individual passages which promised relevant information on a specific case may indeed have been far more important for the way in which such books were generally used than the less linear early modern reading habits on which Mary Fissell has focused in an attempt to apply modern reader response theory to the historical study of popularization.[27] Occasional marginal notes in extant copies point in the same direction. One surviving copy of Tissot's *Avis* only has two short pencil entries (in French), both in the section on diseases of the throat: 'This, I think, is the disease', and four pages later: 'And this is the remedy.'[28] The dozens of pencil marks in another, German copy were also limited exclusively to one section, this time on heat stroke.[29] It is equally striking that hardly any of the approximately eighty copies of Tissot's works I have seen personally (mostly in French and German) carry any pencil marks at all. Many of them look in fact as if they had never been opened and read at all. Modern library holdings of eighteenth-century copies do not necessarily offer a representative sample, but Roy Porter may well have been right in speculating on the 'ceremonial', 'psychological', almost 'talismanic' rather than practical value of owning a Buchan or Tissot.[30]

## Alternative pathways

Medical advice books and similar 'popular' medical writing may have been read somewhat more thoroughly and extensively than the patient letters with their focus on concrete practical advice suggest. But they were undoubtedly only one among several pathways through which expert medical knowledge and practices were communicated to a wider educated public, and it is far from certain that they were the most important one.

To begin with a minor reservation, the boundaries between 'popular' health advice and 'scientific' literature were blurred. What is often called 'popular' medical writing was, in the Ancien Régime, primarily addressed to and read by a small minority only, namely those literate middle and upper classes who could afford and read such books. But many members of these classes were also quite capable of reading the more esoteric, scientific literature of the day and some of them did. When the prescriptions of various country physicians were to no avail, an English vicar near Rugby resorted to the prescriptions in Colbach's book on the mistletoe to cure his son's disease.[31] A number of Hoffmann's patients had previously consulted his voluminous and rather technical *Medicina consultatoria*.[32] One of Tissot's patients diagnosed his disease as a 'gutta serena', after having perused various medical books.[33] On the basis of Pierre Guérin's treatise on eye diseases[34] another patient took his disease, 'without priding himself of judging correctly', to be a 'hemeralopia'.[35] Occasionally patients or their relatives even referred to learned Latin treatises such as Georg W. Wedel's *Tractatus de medicamentorum facultatibus cognoscendis et applicandis* (Jena 1678).[36]

More importantly, the letters suggest that the printed word (or image) in general frequently ranked far behind the oral communication of medical knowledge. Narratives of previous disease episodes, in particular, show an often very extensive mediation of medical knowledge in encounters between physicians and patients. A major reason for this was the configuration of contemporary physician–patient relations. Upper-class patients – hospital inmates are a rather different story – were in a relatively strong position when dealing with a physician. Some patients portrayed themselves as absolutely obedient, whatever the physician declared and prescribed (though we have no means to verify this). But the majority clearly expected to discuss medical matters with their physicians, to defend their own point of view, to reject a diagnosis or a remedy when it did not suit them, or to modify the physician's prescriptions on the basis of their own judgement. Diagnosis and therapy thus were often the outcome of a fairly complex process of negotiation between the sick, their relatives and friends, other healers and the physician.[37] Even famous physicians like Geoffroy, Hoffmann or Tissot had to explain and justify their pronouncements, if they did not want to risk losing their patients. Their letters of response frequently contain a more or less elaborate explanation of the presumable nature of the disease, its effects on the body and the reasons for the choice of a particular treatment.[38] The pressure on the physician was the greater when, as so often, his precarious authority was further jeopardized by the contradictory pronouncements of other physicians on the same case. As a result, in explaining and justifying their own approach, physicians also constantly transmitted and confirmed medical knowledge. Occasionally, this oral communication even seems to have left its marks on a more or less idiosyncratic orthography, when lay persons spelt medical terms wrongly but in such a way that they would be pronounced in more or less the same way: 'hasme' (in French) for 'asthme' for example, or 'cerosités' for 'sérosités', or 'hipercontrie' for 'hypochondrie'.[39]

A second major pathway along which medical knowledge was spread (and constantly reconfirmed) was oral communication among lay people themselves. In many ways, illness was very much a public affair in the Ancien Régime. A quite lively oral exchange of medical information emerges from the patient letters as well as from other contemporary sources, such as testimony in proceedings against irregular healers.[40] People shared their experiences with individual healers or with different types of therapy. They commented on the diseases of others and they gave unsolicited advice. 'Now it being look'd upon as a slight infirmity, amongst my visitant neighbors, everyone is ready to recommend their remedies', the octogenarian John Evelyn complained, when he was badly afflicted with the piles.[41] Similarly, though the sources are more elusive in this respect, ideas about disease causation seem to have circulated, in the form of warnings of the risks involved, say, in exposing the body abruptly to cold drinks or air, or in comments on the reasons of other people's disease or death. Contemporary memoirs frequently contain speculations on the degree to which the death of well-known men and women was self-afflicted by dietetic errors.

All this leads to the conclusion that it was often from a mix of fairly heterogeneous, written and oral sources with frequently contradictory messages that patients and those surrounding them had to construct the most plausible interpretation of the history and symptoms of the disease and deduce the most promising practical approach. Some of the more detailed narratives illustrate this particularly well. When the son of a Swiss priest, for example, was seriously ill, coughing blood and suffering from severe oedema, many people took an interest in his disease. As his father reported, they pressured the parents to call a surgeon and have a fontanelle performed to produce generous suppuration of morbid matter from the swollen limbs. But the parents did not heed their advice and decided to consult Tissot instead. When the patient started vomiting, however, and when little ulcers appeared in his mouth, they consulted a medical dictionary and decided to cool the inflamed stomach with arsenic. And when this had no lasting effect either, they made another, equally unsuccessful attempt with a remedy prepared from woodlouse in wine. This was a treatment which an acquaintance, a priest's wife, had recommended after her own son, as she claimed, had got much better from it, when he suffered from a similar disease. They also tried oxymel with scilla, following Tissot's *Avis*.[42]

## Popularization and medical lay culture

While patients and their relatives frequently indicated where or from whom they learnt about certain drugs or therapies suggested or used in their specific case, the sources of general lay knowledge about the human body and its diseases are much more difficult to pin down. The principles of humoral pathology were so pervasively accepted, among physicians and lay people alike, that they usually cannot be traced to individual popularizing works or a specific mediating agent. It was the kind of knowledge which children already acquired in the course of their socialization, by watching what grown-ups did and hearing what they said. It was part of the dominant cultural repertoire. For this reason, the actual impact of medical popularization and the response to it can best be studied by focusing on fairly recent medical innovations which had not yet become an undisputed part of medical lay culture in general.

As already suggested above, the 'great' new theories, such as the varieties of iatrochemistry, iatromechanism, Stahlian animism or vitalism seem to have had relatively little impact on prevailing lay conceptions of the body and its diseases, however. Only occasionally the letter writers used terms like 'fermentation', 'effervescence' or 'alkali' or they suspected a disturbed 'blood circulation'. And there is no indication at all of a purely mechanical understanding of the human body in the Cartesian sense; instead the patients and those around them seem to have taken it for granted that the body and its inner nature actively and purposefully contributed to the orderly co-operation of the parts and strove to preserve the body from impure, peccant matter, in particular by periodic evacuation. All this does not necessarily imply that patients were unaware of the major systems and schools in contemporary medicine – as we will see the new paradigm of 'nervous sensibility' was accepted very quickly. It rather appears that they simply had no sympathy for those new theories and/or had no use for them. The principles and practices of traditional humoral pathology and Galenic dietetics provided all they needed.

Presumably it was for such reasons that more basic medical innovations (and not just specific new therapies or remedies), to the degree that they can be traced in patient letters and elsewhere at all, were primarily of the somewhat more concrete, medium-range type and usually were at least compatible with traditional humoral-pathological views. For example, the only aspect of 'chemical' medicine which gained wider currency among lay people was the concept of 'acrimonies', which was very close to the popular concept of 'morbid matter' and could easily be absorbed by it. In the rest of this paper, I want to take a closer look at three major, particularly important innovations in eighteenth-century medicine, all concerning such medium-range concepts. Two of them, the medical campaign against masturbation and the rise of the 'nervous diseases', constituted probably the culturally most influential conceptual innovations in Enlightenment medicine. The third, the gradual reinterpretation of menstruation and menopause was, in hindsight, hardly less important but had a rather different fate.

Medical concerns about the physical effects of masturbation grew in the seventeenth century and were increasingly taken up by contemporary moralists writing against 'uncleanliness'. They turned into a regular campaign after the anonymous publication of the famous *Onania or the heinous sin of self-pollution* in 1716.[43] *Onania* went through numerous editions, was plagiarized and translated into German and Dutch. It was in turn the principal source of Tissot's *Onanisme* of 1760,[44] which definitively turned the medical anti-masturbation campaign into a pan-European movement. Excessive semen loss from too much sex had always been considered as dangerous, because semen was thought to be particularly rich in radical moisture, spirits or innate heat or, more generally speaking, in vital, balsamic substances. But *Onania* warned of the even more devastating long-term effects of the local quasi mechanical damage. The 'unnatural' manipulation of genitals was said to lead to impotence, premature ejaculation, infertility and, above all, to chronical genital discharge of gleets or pure semen. This chronic loss of precious matter and not just the semen loss in the act itself weakened the whole body. On top of that, Tissot later pointed out, the nervous system was directly affected by the unnatural arousal and irritation, much more than by the mutual embrace in the matrimonial bed. Numerous testimonials and case histories in later editions of *Onania* and, from 1723, in a supplement, on which Tissot, in turn, largely based his account, illustrated the fatal consequences: impotence, fits and convulsions, severe pain, massive weight loss, anal and urinary incontinence, mood changes and a loss of the intellectual faculties: in short, a horrible physical and mental decline and decay from which death was almost a relief.

Lay fears of the health effects of masturbation can already be traced in letters to Etienne F. Geoffroy in the early eighteenth century.[45] Among the letters to Tissot – an obvious choice for such letters – we then find more than two dozen letters of self-confessed victims of masturbation. They described, sometimes in great detail, a whole range of symptoms and disorders which they had come to attribute to their vice: gleets and impotence, which some patients even accepted as a just punishment of the very parts they had sinned with, but also all kinds of pains and convulsive disorders, weakness and languor, depressed moods or even thoughts of suicide, with symptoms, sometimes worsening in the minutes right after the act. Most of them had years of suffering behind them. Learning about the fatal consequences of masturbation – frequently from Tissot's book – had, as they felt, finally led them to see the truth, and they frequently expressed their gratitude to Tissot.[46]

We do not know how many other patients with somewhat similar symptoms may also have masturbated without accepting this as a possible cause of their suffering. But the letters of those who did as well as an intense public debate make the medical anti-masturbation campaign a striking example of a highly successful dissemination of medical knowledge. What once had been considered as the harmless or even beneficial satisfaction of an urgent natural desire or need, became, for centuries, a major source of fear and guilt. At the same time, the anti-masturbation campaign shows how medical knowledge while it was being mediated served in turn as a powerful mediator of dominant social, political and moral values. The early anti-masturbation literature was quite explicit in this respect. Religious concerns about 'uncleanliness' and the 'murder' of an unborn child combined with 'bourgeois' ideals of self-control and of the importance of marriage and family and with political worries about a decline or degeneration of the population.

The new paradigms of 'nerves' and 'nervous disease', to turn to the second example, had an even greater impact on contemporary somatic lay culture. Among physicians and lay people alike, the 'nerves' developed into one of the leading explanatory models in the eighteenth century.[47] In the learned medical debate, the rise of the 'nerves' was at first driven primarily by a marked shift away from the humours and spirits towards the solids and fibres.[48] From the middle of the eighteenth century this shift was accelerated when the concept of 'nerves' was enriched by new findings on the particular, vital sensibility and irritability of muscles and nerves. Sensibility and irritability, in turn, quickly became characteristics not only of the nervous system but also of the whole body and the whole personality. Since the nerves were distributed throughout the body, excesses or disorders of sensibility or irritability could explain virtually any kind of disease. Some stood out as particularly characteristic, however, such as periodical or intermittent pain, dizziness, cramps, headaches and vague feelings of discomfort, disorders which were all described as rampant.

In patient letters from the early eighteenth century, nerves and nervous diseases only played a very minor role. Symptoms very similar to those later ascribed to the 'nerves' were framed instead by the humoral-pathological concept of 'vapours' or 'vapeurs'. These were usually still understood quite literally as caused by fumes which rose from peccant humours in the belly to the chest, throat and head, causing sensations of tension and tightening, troubling the animal spirits and disturbing their functions. By the late eighteenth century, on the other hand, the language of nerves with its associated concepts of sensibility and irritability was nearly ubiquitous, and when the term 'vapeurs' was still used it usually referred to a 'nervous' affliction. Countless patients now stressed the extraordinary sensibility or irritability of their nervous system. Or they assumed more specifically that their suffering was caused by a desiccation of their nerves, or an excessive tension or contraction, or, on the other hand, an undue loosening and relaxation of their fibres. Going beyond the mere interpretation of their symptoms as nervous, some patients even described a physical sensation of tension, trembling, contraction or pain which they 'felt' in their nerves. In other words, the new medical concepts of 'nerves', irritability and sensibility, were not only widely accepted as a plausible interpretation: they even shaped the very physical experience of illness and, as part of a process of somatic self-fashioning, that of the social 'self' as well.

Like the successful anti-masturbation campaign, the rise of the 'nerves' offers an illustrative example of how newly mediated medical concepts can serve in turn as powerful mediators of social and cultural ideals and norms. The 'language of the nerves' was as fashionable as it was value-laden. It could serve as a somatic idiom of distress, especially for women who had few other means of self-expression. But for the upper classes, having delicate, sensitive nerves and publicly demonstrating them with fainting attacks or other 'nervous' symptoms was also a powerful means of distinction. Their symptoms affirmed that they were down to the innermost parts of their bodies different from those coarse, vulgar lower classes among whom this type of suffering was said to be virtually unknown. Upper-class women, in turn, were seen as particularly prone to 'nervous' symptoms, and the majority of patient letters concerning 'nervous' diseases involved women. This was perfectly in line with most physicians' conviction that the female nerves were particularly delicate and sensitive. Some women almost seemed proud of their uniquely sensitive nervous system, which reacted strongly even to the most minute emotion, the smallest dietetic error, a sip of mineral water. They thus fashioned themselves as the most refined members of society – and literally experienced this refinement in their physical, bodily sensations. But they also confirmed the physicians' claim that their mind and nervous system were less steadfast than that of men, that they were born, thanks to their sensibility, to be their children's loving mothers and their husbands' comfort but unable to take upon them any greater responsibility, let alone a public office.

In medical, academic writing about menstruation and menopause, finally, innovation was somewhat more gradual. For hundreds of years, menstruation had been described as an essential, beneficial evacuation, and the decrease or total suppression of the menstrual flow was considered one of the principal causes of female disease. Enlightened physicians did not entirely give up this idea but they questioned the real importance or necessity of menstruation. Some even argued that menstruation was primarily the product of the unhealthy modern lifestyle of the urban upper classes, which made women accumulate too much blood in their bodies. Country women and women in some exotic countries were said to hardly need any monthly evacuation at all. At the same time, there was an increasing tendency to focus on the negative side-effects or symptoms of menstruation. Menstruation came to be construed above all as a major source of irritation. The irritation got extreme when the process of evacuation was in any way disturbed, so that menstrual blood accumulated or the uterus was activated in vain. The focus of irritation was the uterus, but due to the close links between the uterus, the nervous system and the brain, it was said to spread quickly to the nerves and to the rest of the body. Women who approached menopause and whose periods became irregular were at particular risk. On the other hand, after menopause, once the periods had stopped for good and her body had adjusted to her new state, a woman was said to be often healthier than men of the same age. This was in stark contrast to the fear expressed by Renaissance physicians and occasional later authors that harmful peccant matter would accumulate in the post-menopausal woman, when it could no longer be evacuated every month, causing manifold terrible diseases.

When we look for similar notions among the letters written by sick women, we are in for a little surprise. As I have described elsewhere,[49] most women, and sometimes also male relatives writing for them, mentioned their periods, at least if only to rule out a possible cause of their suffering by stressing that they were normal. They were greatly concerned when they were delayed or totally suppressed or when the amount, colour or consistency changed. But there is hardly any trace of the revaluation of menstruation and menopause which the physicians had proposed. On the contrary: it seems that most women had not even come to accept the interpretation of menstruation which had prevailed among the physicians since around 1600 already. According to most early modern physicians, normal menstruation freed the woman only of the good and pure blood, which she produced in every month. This blood served to nourish the foetus in the case of pregnancy, and when no conception occurred it had to be evacuated only to avoid over-burdening the body and its vessels. Many eighteenth-century women, including those upper-class women represented in the letters, on the other hand, still adhered to the view, once shared by most physicians in the Renaissance: they perceived menstruation as a purifying, cathartic process. It freed and cleansed their bodies from impurities and peccant matter, which, if they were retained would cause all kinds of serious diseases. Accordingly they approached the natural end of menstruation with anxiety. Far into the nineteenth century, as physicians continued to complain, their female patients remained convinced that the end of menstruation would bring all kinds of diseases, including tumours and cancers, due to a local accumulation and hardening of unexcreted impurities. And they sought to prevent such fatal consequences by taking remedies which maintained the menstrual flow or supplemented it, such as drastic purgatives and blood-letting.

In this case then, medical lay culture appears remarkably resilient to the popularization of medical innovation. It is unlikely that women were simply unaware of physicians' reinterpretation. Probably, the traditional notion of menstruation as a cleansing process had become so much part of their cultural repertoire, of their very physical experience, that the new interpretation put forward by physicians simply made no sense to them – after all they 'knew' that menstrual blood neither looked nor smelt quite like the pure blood flowing from the veins of a newly slaughtered animal and they 'felt' that their monthlies freed them from impurity.

## Conclusion

Research on the history of medical popularization must transcend the narrow limits of a textual analysis of the rhetorics and implicit agendas of medical advice books. We also must look for the actual impact of such writing on the medical lay culture rather than assuming that dominant medical discourse will automatically be accepted and assimilated, and we must consider possible alternative means by which medical knowledge was mediated to various parts of the society. Only then will we be able to give meaningful answers to the even more difficult questions as to why different parts of the lay public reacted as they did, why some new medical notions and their propagators were much more successful than others and to what degree this may have been due to the images, values or ideologies which they simultaneously transported. And only then will we be able to identify the ways in which the medical preferences and ideas of lay people may have in turn influenced and shaped the theory and practice of contemporary physicians, that is, to what degree not only individual diagnosis and therapy but also the basic concepts and practices of learned physicians may have been the result of a complex process of negotiation with their patients rather than a mere reflection of scientific 'development'.[50]

## Notes

1. For a good general overview see A. Irwin and B. Wynne (eds), *Misunderstanding Science? The Public Reconstruction of Science and Technology* (Cambridge: Cambridge University Press, 1996); the best introduction to

the history of medical popularisation is still R. Porter (ed.), *The Popularization of Medicine 1650–1850* (London and New York: Routledge, 1992).

2. In addition to the contributions to Porter, *Popularization* see e.g., W. Coleman, 'The People's Health. Medical Themes in Eighteenth-century French Popular Literature', *Bulletin of the History of Medicine*, 51 (1977), 55–74; V. S. Smith, 'Cleanliness: Idea and Practice in Britain, 1770-1850', PhD thesis, University of London (London, 1985); L. Jordanova, 'The Popularization of Medicine: Tissot on Onanism', *Textual Practice*, 1 (1987), 68–79; R. Rey, 'La vulgarisation médicale au XVIIIe siècle: le cas des dictionnaires portatifs de santé', *Revue d'histoire des sciences*, 44 (1991), 413–33; M. Lindemann, '"Aufklärung" and the Health of the People. "Volksschriften" and Medical Advice in Braunschweig–Wolfenbüttel, 1756–1803', in R. Vierhaus (ed.), *Kultur und Gesellschaft in Nordwestdeutschland zur Zeit der Aufklärung* (Tübingen: Niemeyer, 1992), pp. 101–20; C. Verry-Jolivet, 'Les livres de médecine des pauvres aux XVII et XVIIIe siècles. Les débuts de la vulgarisation médicale', in: *Maladies médecines et sociétés. Approches historiques pour le présent*, vol. 1 (Paris, Histoire au présent, 1993), pp. 51–66; D. Teysseire (ed.), *La médecine du peuple de Tissot à Raspail (1750–1850)* (Créteil: Conseil général du Val-de-Marne, Archives départementales, 1994).

3. For a critical assessment of this 'papyrization' see R. Cooter and S. Pumfrey, 'Separate Spheres and Public Places. Reflections on the History of Science Popularization and Science in Popular Culture', *History of Science*, 32 (1994), 237–67.

4. S. A. Tissot, *Avis au peuple sur sa santé* (Lausanne, Grasset, 1761); W. Buchan, *Domestic Medicine, or, a Treatise on the Prevention and Cure of Diseases by Regimen and Simple Medicines* (London,1765); on Tissot see A. Emch-Dériaz, *Tissot: Physician of the Enlightenment* (New York: Lang, 1992); on Buchan see C. E. Rosenberg, 'Medical Text and Social Context: Explaining William Buchan's "Domestic Medicine" ', *Bulletin of the History of Medicine*, 57 (1983), 22–42.

5. J.-A. Helvétius, *Traité des maladies les plus fréquentes, et des remèdes propres à les guérir* (Paris: d'Houry, 1703).

6. Th. Zwinger, *Sicherer und Geschwinder Artzt Oder Neues Artzney-Buch* (Basel: Richter, 1695).

7. H. C. Abel, *Wohlerfahrner Leib-Medicus der Studenten* (Leipzig: Groschuff, 1699).

8. C.-A. Vandermonde, *Dictionaire portatif de santé*, 2 vols (Paris: Vincent, 1760); to my knowledge a thorough survey of this genre has so far been done only for the German literature; see H. Böning, 'Medizinische Volksaufklärung und Öffentlichkeit. Ein Beitrag zur Popularisierung aufklärerischen Gedankengutes und zur Entstehung einer Öffentlichkeit über Gesundheitsfragen. Mit einer Bibliographie medizinischer Volksschriften', *Internationales Archiv für Sozialgeschichte der deutschen Literatur*, 15 (1990), 1–92.

9. I will disregard the many letter consultations written by local physicians and surgeons.

10. On this type of source see G. B. Risse, 'Doctor William Cullen, Physician, Edinburgh. A Consultation Practice in the Eighteenth Century', *Bulletin of the History of Medicine*, 48 (1974), 338–51; L. W. B. Brockliss, 'Consultation by Letter in Early Eighteenth-century Paris: the Medical Practice of Étienne-François Geoffroy', in Ann F. LaBerge (ed.), *French Medical Culture in the Nineteenth Century* (Amsterdam/Atlanta: Rodopi, 1994), 79–117; M. Stolberg, '"Mein äskulapisches Orakel". Patientenbriefe als Quellen einer Kulturgeschichte der Körper- und Krankheitserfahrung im 18. Jahrhundert', *Österreichische Zeitschrift für Geschichtswissenschaften*, 7 (1996), 385–404 (on patient letters to Tissot).

11. L. W. B. Brockliss and C. Jones, *The Medical World of Early Modern France* (Oxford: Clarendon, 1997), p. 283 similarly found 'a basically unitary medical universe' in early modern France.

12. While the theories and discoveries of famous early modern physicians have been studied extensively, we still lack a comprehensive, experience-near historical ethnography of ordinary medical culture and of prevailing views on the body and its diseases; for a brief but useful overview see A. Wear, 'Popularized Ideas of Health and Illness in Seventeenth-century France', *Seventeenth-century French Studies*, 8 (1986), 229–42, and for women's diseases in particular (but based on a physician's case reports) see B. Duden, *The Woman Beneath the Skin: a Doctor's Patients in Eighteenth-century Germany* (Harvard: Harvard University Press, 1991); my own account draws primarily on my current research on patient letters and autobiographical writings, supplemented by physicians' consilia, case histories and similar practice-near sources (cf. M. Stolberg, *Homo patiens. Krankheits- und Körpererfahrung in der Frühen Neuzeit* (Köln: Böhlau, 2003) ).

13. Apart from the *Avis* and his *Onanisme* (see below) Tissot also wrote *De la santé des gens de lettres* (Lausanne: Grasset and Paris: Didot, 1768) and an *Essai sur les maladies des gens du monde* (Lausanne: Grasset, 1770).

14. Bibliothèque Cantonale et Universitaire de Lausanne, Fonds Tissot (henceforth: FT), letter from M. de Pollet, 20 April 1772; translations from non-English sources are mine.

15. FT, letter from F. de Jungken, Kassel, 24 January 1772.

16. FT, letter from M. Le Chartier, 19 January 1776.

17. FT, letter from M. d'Eyrand (?), 6 August 1776.

18. FT, undated letter from J. Gilbert.

19. FT, letter from 30-year-old Amedée Boissière, 3 January 1793.

20. FT, memoir from Vibraye de Roncée, app. letter 10 April 1773.

21. FT, letter from the Marquise, 29 October 1784.

22. FT, letter from the countess, 3 April 1785.

23. FT, letter from Mme de Boubers (?), 30 December 1792.

24. FT, letter from the 50-year-old Abbé de Bonne, 10 October 1774; he suffered from skin disease and feared a cerebral dropsy; FT, letter from Colonel de Juncken, 24 January 1772.

25. *Correspondance de Madame de Graffigny*, 7 vols (Oxford: Voltaire Foundation, 1985–2002).

26. S. Chassagne (ed.), *Une femme d'affaires au XVIIIe siècle. La correspondance de Mme de Maraise, collaboratrice d'Oberkampf* (Toulouse: Privat, 1981),

pp. 105–7, 27 May 1780; in treating members of her family she relied, among others, on Tissot's *Avis*.

27. M. Fissell, 'Readers, Texts, and Contexts. Vernacular Medical Works in Early Modern England', in Porter, *Popularization*, pp. 72–96.
28. Tissot, *Avis*, copy in the Universitätsbibliothek München, shelf mark Med. 621, p. 132 and p. 136.
29. S. A. Tissot, *Anleitung für das Landvolk in Absicht auf seine Gesundheit* (Augsburg/Innsbruck: Wolff, 1772), copy in the Bayerische Staatsbibliothek, shelfmark Path. 1258, pp. 159–68.
30. R. Porter, 'Introduction', in: idem, *Popularization*, pp. 1–16, here p. 9.
31. British Library, Manuscripts department, Ms. Sloane 4075, fol. 56, letter from Mr Davies, 8 July 1728.
32. Friedrich Hoffmann, *Medicina consultatoria*, 12 vols (Halle: Rengersche Buchhandlung, 1721–39).
33. FT, letter from 33-year-old Bruckner from Friesland, 29 November 1789.
34. P. Guérin, *Traité sur les maladies des yeux* (Lyon: Reguilliat, 1769).
35. FT, undated letter from M. de Char(r)itte.
36. Hoffmann, *Medicina consultatoria*, vol. 5 (1726), pp. 326–33, undated letter from a vice-rector.
37. Cf. M. Stolberg, 'La négociation de la thérapie dans la pratique médicale du XVIIIe siècle', in O. Faure (ed.), *Les thérapeutiques: savoirs et usages* (Lyon: Collection Fondation Marcel Merieux, 1999), pp. 357–68.
38. This is particularly striking in É.-F. Geoffroy's quite elaborate answers to often rather brief patient accounts (Bibliothèque Interuniversitaire de Médecine, Paris (henceforth: BIM) Mss. 5241–5). Tissot's responses have survived only in a few cases; otherwise he just left brief diagnostic and therapeutic notes on the original patient letters.
39. BIM 5241 fols 146r–147v, letter from Mme de St Ambroise, spring 1714; FT, letter from the Chev de Bela, 15 March 1773; University Library Erlangen Ms. 1029/I, fols 634–7, letter from Maria Adelheidis S., 5 January 1715.
40. E.g. Hauptstaatsarchiv Stuttgart, A 213, 6734, testimonies in proceedings against a rural female 'doctor', 1773.
41. British Library, Ms. Sloane 4075, fols 94–6, 28 July 1703.
42. FT, series of letters from M. Cart, May–June 1785.
43. *Onania or the heinous sin of self-pollution* (London, [P. Varenne, 1716]); on the historical background and date of publication see M. Stolberg, 'Self-pollution, Moral Reform, and the Venereal Trade. Notes on the Sources and Historical Context of Onania (1716)', *Journal of the History of Sexuality*, 9 (2000), 37–61.
44. S. A. Tissot, *L'onanisme, dissertation sur les maladies produites par la masturbation* (Lausanne: Chapuis, 1760).
45. BIM, 5241, fols 279r–281r, 4 November 1730.
46. For a more detailed account see M. Stolberg, 'An Unmanly Vice. Self-pollution, Anxiety, and the Body in the Eighteenth Century', *Social History of Medicine*, 13 (2000), 1–21.
47. The literature on the 'nerves' and the related issue of 'sensibility' is extensive; useful starting points are G. S. Rousseau, 'Cultural History in a New Key. Towards a Semiotics of the Nerve', in: J. H. Pittock and A. Wear (eds), *Interpretation and Cultural History* (Basingstoke: Macmillan, 1991),

pp. 25–81; G. J. Barker-Benfield, *The Culture of Sensibility: Sex and Society in Eighteenth-century Britain* (Chicago and London: University of Chicago Press, 1992); E. A. Williams, *The Physical and the Moral: Anthropology, Physiology, and Philosophical Medicine in France, 1750–1850* (Cambridge: Cambridge University Press, 1994). For a detailed account of the shift from 'vapours' to 'nerves' in medical lay culture, see Stolberg, *Homo patiens*, 220–60.

48. See on this point R. L. Martensen, 'Alienation and the Production of Strangers. Western Medical Epistemology and the Architectonics of the Body. An Historical Perspective', *Culture, Medicine and Psychiatry*, 19 (1995), 141–82.

49. Cf. M. Stolberg, 'A Woman's Hell? Medical Perceptions of Menopause in Pre-industrial Europe', *Bulletin of the History of Medicine*, 73 (1999), 408–28.

50. Cf. the ground-breaking work by N. D. Jewson, 'Medical Knowledge and the Patronage System in 18th Century England', *Sociology*, 8 (1974), 369–85.

# 7

# Mediating Medicine through Private Letters: the Eighteenth-century Catalan Medical World

*Alfons Zarzoso*

> I'm really pleased with the news of a favourable cure, thank heaven, and I hope the improvement continues. I've relied on it since I read the letters you wrote to Mr Torner and Mr Josef de Capdevila [professors of the Royal College of Surgery at Barcelona] who said it wasn't up to much, for if the pellet had gone more into the eye, you wouldn't have written these letters ... Later, I'll send you the herb you asked me for: if it's not in some apothecary shops, I believe I'll find it in the botanical garden ... Give my regards to the Ladies and cheer Mr Pedro up, whom I've commended to St. Lucy [patron saint of sight].[1]

Jaume Figueras, one of the military men whom Felip Veciana, chief of the Catalan rural police, had in the Barcelona detachment, wrote these concerned words. He explained to his superior, who lived in the village of Valls, fifty-six miles from Barcelona, that he was taking the necessary steps to relieve Felip Veciana's son Pedro of the pain he suffered as a result of a shotgun accident. The words of this letter reveal both mediation in the presence of sickness, and the variety of healing options available in this particular medical world. Moreover, they show the hopes and fears of these people in the face of disease. Similar letters display the same apparent ease of alluding to health in one paragraph and reporting, a few lines later, the current political situation in Barcelona, or the despatch of some clothes, some ounces of chocolate or, perhaps, some 'fine and fresh Flemish butter'. The following lines, written by a member of the Veciana family, echo a strong concern in the face of sickness and place disease in an arena where it was socializsed, understood and discussed. Regarding his father's illness, Ramon Gassol wrote to Felip Veciana that:

> It's eight o'clock and I must tell you that my father had a happier night than we thought, since he slept calmly for five or six hours. That helped to relieve those symptoms that worried Dr Ramon greatly ... Please pass this on to your mother and tell Tereseta to send us, if possible, two pounds of fish and some watermelons; and give her your support, make sure that she is not alone, fearing another disturbance; and ask Dr Pau to visit her frequently.[2]

This chapter argues that we should accept the rationality of individual or collective decisions taken either by those who were sick or by their relatives or friends. It is these actors who are considered the central subject of study and are given priority over sickness and the multiple, diverse resources available which constituted the complex medical world as it was clearly understood by these contemporaries. The existence of a myriad of healing cultures will become apparent by analysing the private letters of an eighteenth-century Catalan family. The letters were written at the end of the eighteenth century to Felip Veciana, the third chief of the Catalan rural police, by his relatives, friends and police subordinates. It is difficult to locate Veciana's personal opinions, voiced in his own words, and it is especially hard to determine his own culture of healing. But this gap can be filled with the help of the replies written by his relatives and it is through their words that we can attempt to reconstruct the medical world of this family.

The use of this source permits the medical historian to overcome or, at least, to complement, the traditional perspective, usually centred on administrative or official records, or the physician's point of view. Thus, such personal sources, which differ from the more partial and biased official records, allow the historian to enter a much more complex medical world. It is in this world where decisions, actors and meanings cannot easily be taken for granted or merely reduced to one exclusive medical perspective. It is now well known that sources such as letters, diaries, accounts and commonplace books permit us to enquire into what men and women actually thought and allow us to understand why they took certain decisions about their health and well-being. Investigating them, it seems clear that there was considerable freedom for action and decision-making as well as language used which remains in the personal, handwritten documents. Yet, as shown at the beginning of this chapter, the knowledge and perceptions of people close to the sick could mediate those acts and decisions. So this chapter will also trace the diverse patchwork of medical knowledge and strategies by explaining how the mediation of these people contributed to expanding the medical resources of this family in a wider, urban context.[3]

Both human and animal medicine should be understood in relation to the medical pluralism characteristic of the Ancien Régime, e.g. from the study of 'those who appeared to sufferers as medically skilled, experienced, or gifted' to relieve them of health problems. This approach is useful as it allows us to obtain a more comprehensive perspective on the medical world as it was perceived by sufferers of ill health. Such a medical world enables us to appreciate both the socio-cultural and economic dimensions that arose at the moment of decision-making before sickness had arisen. Formation of an individual healing culture in the face of disease reveals the existence, overlap and constant interaction of a plethora of available resources, which can be defined through the theoretical model of understanding of three overlapping spheres: popular, medical and religious. The different aspects that feature in the permanent interaction of these spheres reveal a multidirectional appropriation, adaptation and creative elaboration of medical theories, therapeutic remedies, or ritual and symbolic healings. Moreover, they also permit us to address the question of historical change through the study of competition for hegemony that occurs in the worlds of human and animal medicine.[4]

In order to achieve a better understanding of the case study I wish to present, it seems important to place it in its appropriate historical context.[5] It is safe to suggest that in the eighteenth century the Catalan territory was extensively provided with different members of what has been called the 'corporative medical community'. My ongoing research on different historical sources makes the reconstruction of the map of medicine feasible.

From the end of the seventeenth century, Catalonia experienced an extraordinary demographic and economic growth. This steady growth is clearly perceptible in the social organizational changes of the country during the second half of the eighteenth century. The works of Luis García Ballester and Michael R. McVaugh have shown that Catalonia, as part of the Crown of Aragon, demonstrated early acceptance of academic medicine and the control of medical practitioners, thanks to the firm support of religious and royal authorities. Despite the lack of studies on this subject, the evidence so far consulted enables us to assess the growing influence of regular medical practitioners from the late Middle Ages to the Early Modern period. This growth was evident both in numbers and geographical spread in the whole territory, which occurred in parallel with the economic and demographic growth experienced from the end of the seventeenth century.[6]

The eighteenth-century traces confirm the composition of these practitioners in the Catalan area. In this sense, it is correct that surgeons and apothecaries respectively outnumbered physicians in this period. Yet, it is also verifiable, as has been shown in the study of mid-eighteenth-century Andalusian medical practitioners, that the extension of the traditional system of the local council engagement of physicians (known in Catalan as *conducta de comú*) encouraged the institutionalization of healing activities in both rural and urban areas. In the eighteenth century, this way of hiring practitioners was particularly distinguished by the fact that it was promoted for those small towns and villages where it was more difficult to retain a professional. Having the agreement of the majority of the population, physicians, surgeons, apothecaries and veterinary surgeons (known in Catalonia as *menescals*) were tied to these places. Once appointed they were to provide relief for people in return for a salary – paid in cash or in kind such as grain or oil, and sometimes including a house or specific local tax exemptions.[7] On the other hand, authorities encouraged free competition in bigger towns where it was easier for medical practitioners to be required by patients on a regular basis. The study of post-mortem inventories of medical practitioners in eighteenth-century Barcelona reveals great social variety amongst their patients. Thus, such a picture matches up with the importance usually given to population size regarding the number and availability of practitioners. So, it would be wrong to consider this kind of healer, especially physicians, as confined to cities and the exclusive prerogative of the well-to-do.[8]

The proliferation of medical practitioners did not mean, of course, a medical world that was limited only to them. Although not an easy task, it is certainly possible to discover other practitioners besides those who were recognized officially. Despite the difficulties, eighteenth-century Catalan official seriate records allow us to observe the arbitrary policy of authorities, who granted exceptional permission to untrained and irregular healers. There was nothing new in that policy for it had been a constant practice since the late Middle Ages. In the eighteenth century, this policy of exception was not the result of a general lack of trained practitioners, but a consequence of the specific features and services that they offered. On the other hand, competition was not restricted to the medical sphere. The overlapping medical activities practised in the religious sphere have been highlighted through the study of ecclesiastic sources such as sermons or pastoral visits. The religious understanding of disease gave rise to a series of remedies that were put into practice by local priests, who were often denounced to the authorities. In turn, and especially from the Counter-Reformation period, people adopted ritual aspects of the church and transformed them into part of their own cultures of healing. That was added to the extended phenomenon of appropriation and elaboration of the healing properties of the holy pharmacopoeia of sacraments. Thus, medical and religious spheres, in the competition for hegemony, denounced and persecuted the characteristic therapeutics of the popular sphere. The latter informed against the purchase and making-up of specific popular and domestic remedies, which were obtained in the apothecary's shop and were used for the very same medical purpose; and the former reported the improper use of rituals and symbols which were peculiar to the church.[9]

Notwithstanding, official seriate records do not allow us to see the individual or collective rationalisation that remains behind the working of that medical world. Moreover, it is apparent that the use of case histories is the most suitable way to extend the scope beyond the limits of a history based on 'what trained practitioners did'. In fact, what the case study approach highlights is the rationality and pragmatic nature of the choices made by the sufferers from a myriad different healing resources, ranging from self-help to consulting a medical practitioner.[10]

The case history of the Veciana family contributes to understanding such a complex medical world through the words and feelings they expressed in their letters.[11] The Vecianas are not a representative family for writing *history from below* because of their high socio-economic status, but they are, as Andrew Wear would say, 'from the middle, the history of literate but often unimportant people'.[12] The importance of this family in Catalan history derives from the fact that they were the chiefs of the rural police (known as *mossos d'esquadra*) created in Catalonia by the Bourbon dynasty shortly after the end of the War of Succession (1714). Despite having artisan origins, from the social and economic standpoint, the Vecianas became one of the richest families of Valls in the eighteenth century. That was as a result of their position at the head of the police and, as a consequence of this, the perks that the royal power granted them in local government. This situation resulted in a profitable policy of marriage settlements that linked the Vecianas with the owner of the richest local estates. It also led to new family members coming from local physician and apothecary families. The changes in the material culture of the family, the signs of a growing and diversified consumerism, are related to their socio-economic promotion throughout this period, that was demonstrated by their desire for fashionable purchases, specific foods and quality clothes or medicines, which were always expressed in the letters.[13]

Medicine has been considered as another product within the reach of middling consumers and the evidence under study reveals that the Vecianas were not immune to this trend. Their letters show an intense and varied use of different resources available in the medical world described above. In order to enhance the possibilities of this case study, it would be desirable to examine and integrate all of the family sources still in existence. So far, however, the search of their post-mortem inventories has produced no results. The interest in such sources is based on the chance of finding an inventory of books amongst the possessions handed down. As is the case with some other middling consumers, these inventories could prove the existence of medical, advice or self-help books in their library. Books such as the widespread translations of André Tissot's and William Buchan's writings, or even perhaps the possession of a self-medication companion that was sold from 1777 together with a medicine chest, by the Barcelona apothecary Ignasi Francesc Ametller. It is well understood how these popular medical books could have influenced the language employed by middling consumers when describing their health problems. But this is difficult to prove, as it requires a systematic comparison of the sufferers' words with the popular literature.[14] The evidence that remains of the Vecianas shows a rich vocabulary used to describe their diseases. Afflictions are outlined with precision and, interestingly, both medical and vernacular language is used to refer to particular diseases. The words of Francesc Antoni Calbet regretting his son's illness can illustrate this language:

we are again in great distress as our son suffers from epilepsy and spasms. It has happened unexpectedly this morning and he feels very bad; I should not be surprised if he dies today. It is the wish of God and we must resign ourselves to His Divine Will. Please inform my mother and my beloved family and pray to the Almighty and the Blessed Virgin to let him stay with us.[15]

It is also remarkable how these people gave prominence to the description of their feelings of affection, interspersing their opinions with emotions. Such detailed vocabulary and descriptions are basically symptomatic and certainly they could exemplify one of the channels of constructing their own physical world. These features can be reported through the following words of a letter where the sick person shares a particular interpretation of diseases together with both a humoral and symptomatic description of the pain and fear of imminent death. Thus, Mariano Vives wrote from the Barcelona detachment, in the middle of a turbulent period because of the war against the French revolutionaries, that:

> I didn't write to you before because I felt ill … it began in the head, where all diseases begin, but I still feel very bad. I'm not in bed because I have a hot temper. I'm only in bed when I can't bear it any more. But now, I'm alone, far away from home, with only the help of God's love, Our Lord; and I think that if it lasts long I'll die, as I'm suffering from a continuous fever.[16]

It was also in the context of the war against the French Revolution that some letters placed specific blame for disease. One of these letters firmly expressed the causes of disease in a besieged town close to Barcelona by stating that, 'there is a rumour going round that the French had poisoned the waters of the village'.[17] These words implied an actual explanation for people's deaths, with water being the means of spreading that disease, even though the presence of the French was doubtful. Clearly, such an explanation can be interpreted as a reaction in the face of a disruptive situation in which disorder even reached traditionally socially stigmatized institutions such as the General Hospital of Barcelona. In fact, successive deaths of the representatives of the social order, such as the councillors of the municipal government of Barcelona, military authorities, the local nobility and the city bishop, contributed to the association of disease with the 'other', the French. As in the not too distant past, when the plague had visited Barcelona with distressing results, people tried to blame those considered a threat to the established social order. Thus, the daily report of the current political situation in Barcelona, written by Mariano Vives to his commandant, plainly illustrates just such a pitiful state of affairs:

> In Barcelona, there are a lot of sick people as a result of the evil nature of the French; in the meanwhile, when people enter the

hospitals, everybody, physicians, surgeons, and assistants, fall ill and soon die.[18]

It must also be emphasized that these people possessed abundant resources. Thus, the presence of representatives of the corporative medical community should not obscure the existence of other resources, treatments and remedies administered without the supervision of any medical practitioner. Again, it would be desirable to find out how all the family strategies had interacted. Yet the search of the personal accounts, dietaries and family manuscript recipe books which Núria Sales has employed to describe some of the Vecianas' medical usage and customs has also been unsuccessful. According to her descriptions, these sources are valuable as they contain, amongst mainly culinary recipes, an arsenal of popular therapeutics, household remedies and medical prescriptions. A stock of useful medical knowledge in the hands of these people before they became ill was transformed into useful means of self-diagnosis and self-medication. The description of remedies and cures that can be found contained in those books reveals a popular therapeutics devoted to the healing of a wide catalogue of minor diseases, those that did not require the presence or supervision of a trained medical practitioner. So, they recorded lay remedies for headaches, toothaches, constipation, stomach disorders such as diarrhoea, coughing and chest pain, common colds and fevers, and also more sophisticated prescriptions for major diseases. Among such practical knowledge there were also recipes for food and, interestingly, specific formulas for cosmetic products of every sort, such as face creams, potions to treat bad breath and strained eyes, poultices to retard incipient moustaches, or different colourings, oils and waxes for the lips, face and skin. Whether or not these remedies were efficient, I think all of these aspects taken together are significant for an understanding of the great complexity of that medical world. For they show how individuals were often operating quite independently of medical practitioners by their possession of, or obtaining through exchange, an extraordinary array of traditional resources.[19]

Self-medication and the resource of advice from relatives and friends appear to be the basis of the Vecianas' medical world and it shows the great control these people had over the treatment of certain diseases. In fact, the letters frequently permit us to see how relatives put their own experience to use in order to make recommendations for treatment. Thus, in some letters the importance of personal experience of an illness is emphasized when home remedies are shared or a particular treatment is recommended as one that has worked in the past. Extensive bits of advice were intersected into letters; these were given without pretension to impose any kind of authority but to offer help through particular experiences and popular wisdom. For instance, this was the case with a letter written by Josep Brines, who said that:

> I've just met a man who suffered from the same pain that distresses your son Pedro, and he's now recovered from it. He assures me that the only and most effective remedy is dried almond-shell water, which must be boiled in natural water … And later he'll be relieved and perhaps will attain a complete recovery, since the person who has tried this remedy is now entirely free of that pain.[20]

As has been pointed out, letters also express the control sufferers had over their treatment and this is clear when they state their determination to change a therapy by arguing that it could be more economical, beneficial or simply more comfortable. By acting as their own physicians, these people exerted control over their bodies and diagnosed themselves according to their own experience. Perception of the self is often based on an understanding of the traditional humoralism and, as a result, the vocabulary used can be conceptualized as the practice of one's own self-regime. This view becomes apparent in some letters where friends advised the Vecianas about the importance of following certain health rules in order to restore balance: avoiding anger and embracing merriment and laughter instead, walking every day, drinking water, eating seasonal fruit and green vegetables and avoiding salted and spicy food. Such an appropriation and re-elaboration of the prevailing medical discourse, together with a firm resolution of self-control, appear clearly in a letter written by a family friend who said that:

> I'm determined not to take the waters of La Espluga this year as I've found something similar but cheaper and more comfortable. It consists of a glass of fresh water with a few drops of vitriol and chocolate … I've been taking it for several months and it has been of great benefit. I don't need to take the waters. I would be happy if you recover from the weakness caused by bleedings. Now is a good season to bring about remedies and you should eat vegetables and fruit.[21]

These features can also be observed in moments of transition related to the meaning of life, such as births or deaths. It becomes almost a commonplace to note how the writers of these letters felt an interest in the health of the family or stated their own health concerns in a routine way. Concerns of both writers and recipients were related in such a way as to confirm the idea of the existence of a first step in the building of cultures of healing based upon shared advice, feelings, hopes and fears. For instance, the announcement of a birth entailed grounds for joy but also could be an occasion for concern regarding the state of health of both mother and newborn child. Francesc Antoni Calbet was so happy that he considered it necessary to express his satisfaction together with his concern for the health of his father-in-law, Felip Veciana:

> I inform you of the happy delivery of Marianita, who has plenty of milk and has been feeding our baby since yesterday; and both together feel perfectly sound, by the grace of God, and Marianita looks as if she hadn't had a baby. I'm very glad to know that Father is getting better; he should go for a walk every day and be merry as we are very concerned about him.[22]

Narratives of friends and relatives were almost always the first indication of a sickness and, consequently, through their words, the description of their state of mind. A disease that ended in death was indeed an important reason for writing a letter. Some of the surviving examples show how those who were mourning wrote their letters in tears. Yet, that bitter situation did not prevent them from describing, one might say warning of, the tragic progress of a disease that was, presumably, erroneously diagnosed. Those were the feelings of Teresa Tort, who wrote to her uncle, Felip Veciana, in a depressed state that:

> his illness began with a cold and the physician was convinced that it was certainly a catarrh for the next seven days. Then the pain became so acute that after three days he died. You, Uncle, can imagine my sorrow ... as we thought it was a cold and now we feel rather alone. You can realise my tough convalescence ... I can assure you, uncle, that words fail me to express my sorrow and pain. If God doesn't take good care of me I won't succeed in this battle.[23]

Disease and death motivated these people to turn to God. As a result, there were continual references to an intense religiosity when they fell sick. Religion played a key role in the face of death and was a part of the overlapping resources available. Thus, entrusting oneself to God or to Holy Mary or praying to the holiness of a shrine was not incompatible with other treatments, but part of the variety that shaped the rationalization of every act within their culture of healing. There was a clear link between health and holiness and this was plainly stated through their words together with the description of a variety of resources utilized. After having arrived at his home town, corporal Jaume Figueras wrote to his commandant that,

> I didn't write to you before because the whole family fell ill and I was the one who got worse, arriving at the doors of death. So we consulted the doctors of Hostalrich and with the help of the Almighty and the Blessed Virgin of Puig de Bellver we are now getting better ... all the more so given the great epidemic that razes this area ... I was given permission to come to my brother's rectory in order to convalesce and recover completely ... I arrived exhausted and because of my weakness I'm afraid it will take me a long time to recover; yet, being at home, in a land like the Holy Mount Tabor, I hope to get better quickly with the help of the Almighty and the remedy of our Liberata [his wife], who is still suffering from fevers ['tercianas'] which has forced her to give the baby she had to a foster-mother.[24]

There was also a relationship between sickness and sin. Some letters allow us to observe that these people accepted sickness and death as a fact of Divine Will and openly stated this in describing their condition as sinners. In turn, this faith helped them to accept death and to become reconciled to God, which fortified them. Francesc Antoni Calbet's narrative about his daughter's death reveals a Christian understanding of it and is also a clearly detailed and medicalized way of confronting sickness from a lay point of view. He wrote to Felip Veciana that:

> finally yesterday ... God called for the soul of our eldest daughter, who had been suffering for three weeks ... That sickness was mainly a strong and continuous colic originating in very hot humours and an entire corruption. She had a continuous bowel movement every day and I can assure you that there was an unbearable odour coming from her room ... We must resign ourselves to the Will of God, who, I don't doubt it, has determined to take her away. Otherwise He would have granted our request, although we are nothing but sinners. That's enough! I'm so confused that I don't know what I'm writing.[25]

There are a number of letters that enable us to observe even more clearly how mediation in medicine was achieved and also how the three spheres described above overlapped from the standpoint of the Veciana family. What is remarkable about the following examples is the distance that separated the Vecianas from the growing city of Barcelona. Seemingly, it could be argued that such a distance divided two different worlds regarding the number and kind of medical resources available; which is true in part. Yet, in spite of the fact that the Vecianas lived in the provinces, their home town was adequately provided with midwives, barbers, surgeons, apothecaries and physicians. Actually, Valls, their hometown and headquarters of the rural police, experienced a remarkable growth of population through the century, from around 3000 inhabitants in 1718 to nearly 8000 inhabitants according to the census of 1787, and the number of medical practitioners grew comparably.[26]

The Vecianas enjoyed high social status and were affluent enough to be able to afford every sort of medical service on a regular basis. Moreover, they had family relationships with medical practitioners. Felip Veciana's mother, Francesca Dosset Gassol, came from a well-connected family of Valls apothecaries. He also had a sister, Maria, who married a physician from Valls, Magí Vives. Once Felip Veciana became the head of the family, they entrusted their supply of medicines to a local family of apothecaries, the Carrera family. In 1791, the widow Carrera demanded that the Vecianas fulfil the obligation they had contracted a long time ago with that apothecary to be supplied with medicines and to settle the unpaid account of around 870 Catalan pounds. This was a characteristic case listed in the Catalan credit economy which was based on postponed debts in order to ensure a regular relationship with customers.[27] On the other hand, the existence of a lawsuit between the Vecianas and the medical doctors Antoni and Albert Güell, father and son respectively, allows us to confirm that they did use local doctors alongside other medical resources. The Güells tended the Vecianas for many years, between 1774 and 1790 at least. The Güells sued because of work carried out through this period that remained unpaid. According to the doctors' accounts, the family called on them on several occasions. It seems that the Vecianas sought their professional help intensively, since the Güells attended family members and servants regularly visiting them day and night in an extraordinary way; they also consulted other doctors more than 250 times about problems that were not considered part of their medical relationship. Like the apothecary Carrera, the Güells operated a credit system; according to the lawsuit it was normal to settle debts from time to time, but never definitively, for the engagement was based on reciprocal trust.[28]

Despite having and relying on a variety of trained medical practitioners that operated in their home town, the Vecianas requested specific services of their friends and subordinates from Barcelona. The improvement of the road system in the eighteenth century made the satisfaction of such demands easy and contributed to consolidating the medical culture of the Vecianas. Once the Vecianas decided to break their medical engagement with the Güells, Dr Ramon Pons, a young physician recently graduated in 1791, looked after them. It was through the letters that Felip Veciana sent to his men in Barcelona that Dr Pons made medical requests, such as for specific compounds that were difficult to obtain in a small town. Although the requests were made personally by Dr Pons, the written replies were never addressed to him, but to Felip Veciana. For instance, Mariano Vives explained to Felip Veciana his journey from one apothecary's shop to another in order to find a specified amount of a medicine, called Rose of Genoa, that had been requested by Dr Pons. Finally, it was a well-known physician and apothecary of Barcelona, Dr Benet Pujol, an acquaintance of Dr Pons, who compounded the medicine that was then despatched to Valls.[29]

In the spring of 1796, Felip Veciana caught a serious illness that slowly brought about his death in 1798. Letters make it possible to follow the route that led the family from one type of care to the next. In fact, letters make it plain how family and friends looked for the best solutions, made recommendations, commented on the treatments, and worried about his health. Certainly, friends and subordinates took the initiative and sought to help him as far as they could from their vantage point at Barcelona. Despite the existence of a number of physicians in the city, they decided to consult Geremí Clarós, a physician who practised in the village of Horta, near Barcelona. In the postscript to a letter it can be read that, 'yesterday, we went to Horta ... to meet a physician of reputation and explained to him your ills so that he could relieve you ... since he has healed a lot of people'.[30] A few days later, they wrote another letter that enclosed '... a note written by the physician of Horta so that you could deliver it to Dr Pons, who should send me his opinion; and this physician insists on healing you and remember that health is our first duty'.[31] From then onwards there began an exchange of medical reports mediated through the letters of the police subordinates. Dr Geremí Clarós wrote a report stating his opinion on the several afflictions of Felip Veciana, their causes and a detailed catalogue of remedies for every one of these pains.[32] Mediating medicine was apparently here reduced to the transmission of letters. Yet if those remedies did not have the desired effect, the Veciana men made use of the medical report written by Dr Pons to search for another treatment. Interestingly, they trusted in the personal experience of a local weaver, Miquel Arno, who read, decoded and interpreted Dr Pons' medical report and eventually prescribed and wrote a note regarding the remedy, which he thought 'will heal you with God's favour'.[33]

Mediating medicine acquires here a particular meaning since these people established a relationship with a lay practitioner that was based on the acknowledgement of authority, a medical authority, but it was not subject to the power of one over the other. There were two main mediators in this affair. One of them was the corporal Jaume Navarro, who apparently had already taken the weaver's concoction. He wrote an encouraging letter to Felip Veciana outlining the advantages the remedy would bring him:

> I'm sorry that you're not feeling quite yourself. Pray to God and trust him, as He would give you back your health if that were suitable to you. I like what you say about the remedy of Miquel Arno, the silk weaver ... he promises that regarding scurvy you will be relieved, with God's help, and otherwise he affirms that the remedy will by no means harm you. I know that it is a bit tart, but be patient.[34]

The other mediator was not a police subordinate but a priest, Francesc Martí, who also encouraged Felip Veciana to take the potion. In contrast to the general medical pronouncement of the day that reliance on folk remedies befitted only peasants or superstitious minds, and contrary to the view of some historians who have contended that folk medicine and self-medication were peculiar to the popular or working classes, this case shows the opposite – that lay people when sick did not entrust their bodies solely to medical practitioners but to the whole array of medical resources available. As a matter of fact, it is worth echoing the voice of the priest, who assured Felip Veciana about the harmless properties of the potion, described its benefits and explained to him how it must be taken:

> I've just met Miquel Arno, the weaver, who says that the remedy won't harm you. You must gargle with it every day after dinner. It will strengthen your teeth, relieve your pain and preserve you from

sores. If you don't want to feel ill you should drink hot water at all times, not cold water, so you'll heat your stomach and get better. He knows it from experience, as when he was a dropsical man, suffering from scurvy and rejected by physicians he treated himself with hot water and he's now perfectly sound ... It doesn't matter if time passes, the remedy is still good and keeps the same virtues. Drinking hot water, he says, will take away that unquenchable thirst, but not cool water, only tepid water, as he recovered and since then he's always been sound.[35]

Deprived of hope, Felip Veciana died in March 1798. Yet, before dying there was still room for a consolatory spiritual death. In this respect, the case could epitomize what has been highlighted as the importance of observing 'responses to health and sickness as constitutive parts of whole cultural sets'.[36] To this effect, in a deeply religious cultural context and given the ineffectiveness of medical healing at that time, perhaps spiritual healing was the best remedy Felip Veciana could obtain, as a friend exhorted him:

Do try to amuse yourself and recover and let matters ride because God allows prosperity for bad people with two different aims: one is to try good people and the other is to inflict on the wretched the worst troubles; on the contrary, God allows mishaps for good people to try the greatest utility and profit from their souls. Such a thought is what must encourage you to suffer with Christian patience and resignation the troubles inflicted on you by the Holy Hand.[37]

In conclusion, the overall picture shows that examining the history of medicine from the sick person's point of view enables us to observe a complex medical world in the eighteenth-century Catalan context. If we want to evoke this historical category we need to broaden the scope of our enquiry and, as a result, imagine the medical world as rich and diversified, just as were the motivations of those in the past who searched for a remedy and those who dispensed medicine. A comprehensive history of the medical world in early modern Spain has still to be written. In this chapter I have tried to demonstrate that the key to the whole argument is the use of different historical sources, especially those that allow the sick to speak. It has been pointed out that it is difficult to integrate the words of sufferers and specific cases into a broader social and political structure, but it is desirable to go beyond a traditional history limited to the voices of trained practitioners. In this respect, it has been shown that despite the *longue durée* of religious and humoralist explanations of disease, their ambiguity and uncertain efficacy gave the sick the freedom to build their own cultures of healing, which in turn came about in specific contexts of negotiation, regulation and repression.[38] Although this chapter cannot offer definitive conclusions, it does illustrate how a careful interpretation of the Vecianas' correspondence enables us to reconstruct the medical world as it appeared to them. While this case study is not necessarily representative, it does reveal clearly how the sufferers retained control over remedies and resources. It also shows a particular use of language, especially medical language, related to their understanding of their bodies and the diseases that affected them. The importance of religion in such a cultural context is beyond doubt and this was continually stated in their letters. On the other hand, mediating medicine appears of paramount importance in the search for remedies. In fact, personal experience and the acknowledgement of authority in people who were not regarded as trained practitioners presumably made that medical world a place free from the inequalities peculiar to those relationships, based on the idea of power, between lay people and medical practitioners.

## Notes

I would like to thank Jon Arrizabalaga, Josep Pardo Tomás, Enrique Perdiguero and Núria Sales for their helpful advice.

1. The material used here comes from the collection of more than a thousand family letters edited by A. Borruel, *Els mossos d'Esquadra: aportació documental a la seva història (1741–1821)* (Valls: Institut d'Estudis Vallencs, 1998). Unless otherwise stated, all the source references come from this book that hereafter will be cited as ME. As all letters were written in Catalan or Spanish, I am responsible for their translation into the English language. Letter dated at Barcelona, 10 December 1796, ME, p. 440.
2. Letter dated at Montblanc, 18 September 1794, ME, p. 262.
3. On the use of these kinds of sources, see S. King and A. Weaver, 'Lives in Many Hands: the Medical Landscape in Lancashire, 1700–1820', *Medical History*, 45 (2000), 173–200.
4. See R. Porter (ed.), *Patients and Practitioners: Lay Perceptions of Medicine in Pre-industrial Society* (Cambridge: Cambridge University Press, 1983) and R. Porter, 'Doing Medical History from Below', *Theory and Society*, 14 (1985), 175–98. On the concept of medical world, see L. Brockliss and C. Jones, *The Medical World of Early Modern France* (Oxford: Clarendon, 1997), pp. 1–33. On the model of the three spheres, see D. Gentilcore, *Healers and Healing in Early Modern Italy* (Manchester/New York: Manchester University Press, 1998).

5. See the works of E. P. Thompson, 'Anthropology and the Discipline of Historical Context', *Midland History*, 3 (1972), 41–55; 'Folklore, Anthropology and Social History', *Indian Historical Review*, 3 (1978) 247–66.

6. See M. R. McVaugh, *Medicine Before the Plague: Practitioners and their Patients in the Crown of Aragon, 1285–1345* (Cambridge: Cambridge University Press, 1993); C. Ferragud, 'Els professionals de la medicina (físics, cirurgians, apotecaris, barbers, i menescals) a la Corona d'Aragó després de la Pesta Negra (1350–1410): activitat econòmica, política i social' (unpublished PhD thesis, University of Valencia, 2002).

7. On the local engagements of practitioners, see Aragon Crown Archive, Royal Audience, Consultes, book 158, 145v–146v, Report of the Royal Audience to the Castilla Council, Barcelona, 25 January 1737.

8. The books of *Cartes* and *Consultes* of the Royal Audience between 1716 and 1821 (Aragon Crown Archive), the books of *Sanidad* of the Barcelona Town Council in the same period (Historical Archive of Barcelona) and the study of a number of eighteenth-century post-mortem inventories (Archive of the Notarial Protocols of Barcelona) allow me to point out the ideas stated. See also A. Zarzoso, 'Protomedicato y boticarios en la Barcelona del siglo XVIII', *Dynamis*, 16 (1996), 151–71; X. Sorní, 'Notas sobre conductas de boticarios en poblaciones catalanas a mediados del siglo XVIII', *Bol. Soc. Esp. Hist. Farm.*, 38 (1987), 219–27 and 'Notes sobre conductes mèdiques catalanes pels volts de 1750', *Gimbernat*, 18 (1992), 157–67; M. Domènec, 'Estructura geogràfica de les botigues d'apotecari en la Catalunya dels segles XVII i XVIII', in J. M. Camarasa et al. (eds), *I Trobades d'Història de la Ciència i de la Tècnica* (Barcelona: Societat Catalana d'Història de la Ciència i de la Tècnica, 1994), pp. 203–25; T. Ortiz, C. Quesada, J. Valenzuela, M. Astrain, 'Health Professionals in Mid-eighteenth-century Andalusia: Socio-economic Profiles and Distribution in the Kingdom of Granada', in J. Woodward and R. Jütte (eds), *Coping with Sickness: Historical Aspects of Health Care in a European perspective* (Sheffield: European Association for the History of Medicine and Health, 1995), pp. 19–44.

9. See R. Jordi, *Aportació a la història de la farmàcia catalana (1285–1997)* (Barcelona: Fundació Uriach 1838, 1997); M. Gelabertó, 'Religión, enfermedad y medicina popular en la Cataluña del siglo XVIII', *Historia Social*, 26 (1996), 3–18; M. Camps and M. Camps, *Santuaris lleidatans amb tradició mèdica* (Lleida: Seminari Pere Mata-Universitat de Barcelona, 1981); J. Martí, 'Medicina popular religiosa a través dels goigs', *Arxiu d'Etnografia de Catalunya*, 7 (1989), 171–203 and *La medicina popular catalana* (Barcelona: Labor, 1992); F. Ribas, *Els goigs de l'Hospital General de Santa Creu de Barcelona* (Barcelona: Seminari Pere Mata-Universitat de Barcelona, 1994); A. Zarzoso, 'La societat davant els animals: ramaderia i menescalia a la Catalunya del segle XVIII', in *IV Trobades d'Història de la Ciència i de la Tècnica* (Vic: Societat Catalana d'Història de la Ciència i de la Tècnica, 2000), pp. 107–11; M. L. López Terrada, 'El pluralismo médico en la Valencia foral. Un ejemplo de curanderismo', *Estudis*, 20 (1994), 167–81; E. Perdiguero, 'Protomedicato y curanderismo', *Dynamis*, 16 (1996), 91–108; M. J. Merinero, *Percepción social de la enfermedad en tiempos de la Ilustración* (Cáceres: Universidad de Extremadura, 1995).

10. Sources such as private diaries or popular books permit one to examine the perception of both individual and collective sickness. See F. Sardet, 'Le partage des savoirs. Aux sources d'une sociologie historique de la santé à Genève, XVIIe–XVIIIe siècles', in *Maladies, Médecines et Sociétés* (Paris, l'Harmattan, 1993), pp. 41–50; J. S. Amelang, *A Journal of the Plague Year: the Diary of the Barcelona Tanner Miquel Parets, 1651* (Oxford: Oxford University Press, 1991); M. Lindemann, *Health and Healing in Eighteenth-century Germany* (Baltimore: Johns Hopkins University Press, 1996).

11. See R. Porter, 'The Patient in England, c.1660–c.1800', in A. Wear (ed.), *Medicine in Society: Historical Essays* (Cambridge: Cambridge University Press, 1996), pp. 91–118; and A. Borruel, 'Algunes qüestions entorn la mort del comandant de les esquadres', *Gimbernat*, 17 (1992), 133–9.

12. See A. Wear, 'Interfaces: Perceptions of Health and Illness in Early Modern England', in R. Porter and A. Wear (eds), *Problems and Methods in the History of Medicine* (London: Croom Helm, 1987), pp. 230–55.

13. On the Vecianas, see N. Sales, *Història dels mossos d'esquadra. La dinastia Veciana i la policia catalana del segle XVIII* (Barcelona: Aedos, 1962) and *Una vila catalana del segle XVIII* (Barcelona: Dalmau, 1962); A. Borruel, *Les esquadres de Catalunya a finals del segle XVIII: de la prosperitat a la decadència* (Valls: Institut d'Estudis Vallencs, 1994).

14. See E. Perdiguero, 'The Popularization of Medicine during the Spanish Enlightenment', in R. Porter (ed.), *The Popularization of Medicine, 1650–1850* (London and New York: Routledge, 1992), pp. 160–93; R. Porter, 'Spreading Medical Enlightenment: the Popularization of Medicine in Georgian England and its Paradoxes', in *The Popularization of Medicine*, pp. 215–31; M. Fissell, 'Readers, Texts, and Contexts: Vernacular Medical Works in Early Modern England', in *The Popularization of Medicine*, pp. 72–96; R. Jordi, 'Manual de automedicación del boticario barcelonés Ignacio Francisco Ametller, segle XVIII', *Butlletí de la Societat d'Amics de la Història i de la Ciència Farmacèutica Catalana*, 11 (1996), 12–32.

15. Letter dated at Tarragona, 5 July 1792, ME, p. 111. Francesc Calbet to his father-in-law, Felip Veciana.

16. Letter dated at Barcelona, 1 March 1794, ME, pp. 230–1. Mariano Vives to his uncle, Felip Veciana.

17. Letter dated at Vespella, 24 September 1793, ME, pp. 189–90. Corporal Figueras to Felip Veciana.

18. See D. Nelkin and S. L. Gilman, 'Placing Blame for Devastating Disease', in A. Mack (ed.), *In Time of Plague: the History and Social Consequences of Lethal Epidemic Diseases* (New York and London: New York University Press, 1991), pp. 39–56. Letter dated at Barcelona, 11 February 1794, ME, pp. 223–4. Mariano Vives to Felip Veciana.

19. On this see N. Sales, op. cit. note 13, pp. 21–7. I would like to thank Núria Sales, Anna Borruel, Salvador Ramon and Antoni de Veciana for their help with my research material. An important part of the Veciana papers is now dispersed and in the hands of several family descendants.

20. Letter dated at Tarragona, 25 October 1800, ME, p. 592. Josep Brines to Maria de Miró, Felip Veciana's wife.

21. Letter dated at Barcelona, 28 May 1796, ME, pp. 365–6. Francesc Martí, priest, to Felip Veciana.

22. Letter dated at Tarragona, 13 August 1796, ME, p. 414. Francesc Calbet to Felip Veciana.

23. Letter dated at Balaguer, 23 January 1793, ME, pp. 142–3. Teresa Tort to her uncle, Felip Veciana.

24. Letter dated at Vespella, 24 September 1793, ME, pp. 189–90. Corporal Figueras to Felip Veciana.

25. Letter dated at Tarragona, 20 July 1794, ME, p. 251. Francesc to Felip Veciana.

26. See P. Vilar, *Catalunya dins l'Espanya Moderna* (Barcelona: Curial, 1988), vol. II; M. Domènec, 'Estructura geogràfica'.

27. Letter dated at Valls, 4 August 1791, ME, pp.65–6. Josepa Carrera Ortega to Felip Veciana.

28. Letter dated at Valls, 20 January 1792, ME, pp. 79–80. Report on the works carried out by the physicians Dr Antonio Güell and Dr Albert Güell. Letter dated at Valls, 9 April 1792, ME, p. 96. Felip Veciana to Dr Albert Güell. Letter dated at Valls, 15 April 1792, ME, pp. 96–7. Dr Albert Güell to Felip Veciana.

29. Letter dated at Barcelona, 8 March 1794, ME, pp. 232–3. Mariano Vives to Felip Veciana.

30. Letter dated at Barcelona, 2 July 1796, ME, pp. 390–1. Mariano Vices to Felip Veciana.

31. Letter dated at Barcelona, 13 July 1796, ME, pp. 401–2. Mariano Vices to Felip Veciana.

32. Letter dated at Barcelona, 23 August 1796, ME, pp. 415–16. Corporal Jaume Navarro to Pere Màrtir Veciana.

33. Letter dated at Barcelona, 30 August 1796, ME, p. 417. Corporal Jaume Navarro to Felip Veciana.

34. Letter dated at Barcelona, 30 September 1796, ME, p. 425. Corporal Jaume Navarro to Felip Veciana.

35. Letter dated at Barcelona, 15 October 1796, ME, p. 427. Francesc Martí, priest, to Felip Veciana.

36. See Porter, *Patients and Practitioners*.

37. Letter dated at Barcelona, 11 November 1797, ME, p. 514. Josep Constans, friar, to Felip Veciana.

38. Wear, 'Interfaces'.

# 8

# 'Mediating Sexual Difference': the Medical Understanding of Human Hermaphrodites in Eighteenth-century England

*Palmira Fontes da Costa*

> They are not two, yet the form is twofold;
> We can not say, that it is the body of a boy or a girl;
> They are neither, though they appear to be both.
>
> Ovid, *Metamorphoses*

Should human hermaphrodites be considered the 'true' embodiment of a mingled male and female nature or, instead, should their sexual duality be ascribed to misleading appearances? This chapter documents one critical moment in the history of sexual identity in mid-eighteenth-century England when this question became pertinent and controversial for the medical profession. The discussion of hermaphroditism offers a unique window into the complex web of mediations involved in the making and diffusion of medical knowledge

Previous to the eighteenth century, the medical literature was not in agreement on the causes, classification and status of hermaphrodites, but their existence, in general, was not questioned.[1] Instead, in a period where both men, women and hermaphrodites were understood to exist, the important question had been who was to decide the hermaphrodite's predominant sex and on what grounds. Then, as now, establishing the hermaphrodite's gender was vital since it affected its legal and social position. Once typed as male or female by medical and legal authorities, the hermaphrodite was entitled, with a few exceptions, to all the prerogatives of that sex.[2] The legal offence was to fail adhering to the established gender identity by claiming the prerogatives of both genders. Furthermore, if hermaphrodites moved back and forth between genders, their sexual relations could be stigmatized as sodomy.[3]

As Lorraine Daston and Katherine Park have remarked, the understanding of hermaphrodites in this period was influenced by two distinct and, in many ways, contradictory traditions, the Hippocratic and the Aristotelian.[4] On the one hand, in the Hippocratic tradition sex existed along a continuum from extreme male to extreme female and hermaphrodites were considered beings with a truly intermediate sexual nature.[5] On the other hand, within the Aristotelian interpretation of sexual difference, male and female corresponded to polar opposites admitting no meaningful intermediate states and hermaphrodites were viewed as beings with doubled or redundant genitalia.[6] However, Daston and Park have pointed out that the sixteenth- and seventeenth-century literature on this subject often provided an uneasy synthesis of these two antagonistic approaches.[7]

In comparison with the previous period, eighteenth-century medical debates on hermaphrodites represented an effort to clarify and unify the interpretation of sexual nature. In this chapter, I will show that the most radical means used to achieve this aim was to erase the concept and the figure of the hermaphrodite altogether from the medical understanding of sexual difference. To explain away hermaphrodites as the result of misleading appearances meant discarding the Hippocratic model and definitely embracing the Aristotelian, which presented the ambiguity of hermaphrodites due only to a superficial and meaningless local excess of matter. At the same time, the removal of the figure of the hermaphrodite enabled the consolidation of a binary understanding of sexual difference which had crucial implications for the moral and social order. I will also show that eighteenth-century medical debates concerning the nature of human hermaphrodites were not confined to which traits and which body parts would count as essentially or significantly masculine or feminine. Also at stake were more general issues concerning the nature and value of anatomical evidence. Two of these were how and by whom appearances could be effectively distinguished from the reality of nature and what the specific role of visual representations should be in the construction of anatomical knowledge. Furthermore, the eighteenth century was a period with a competitive market in medical practice, and the discussion about hermaphrodites involved negotiations among medical men concerning their professional authority and expertise.[8] In large part, this was due to the fact that the study of hermaphrodites, as well as of other monstrous beings with an ambiguous anatomy, provided a good opportunity for physicians and surgeons to exhibit their anatomical skills and knowledge.

The analysis presented in this chapter will be focused primarily on a famous African hermaphrodite exhibited in London throughout 1741. This case will be used as the standpoint from which to analyse the complexity of approaches and interests involved in the perception of hermaphrodites in mid-eighteenth-century England. I will undertake my analysis in three stages. First, I will focus on the settings and audience of the exhibition of the Famous African. Then, I will consider the meaning of visual representations in the understanding of human hermaphrodites. Finally, I will address some of the moral and social implications of eighteenth-century medical views about the nature of human hermaphrodites.

## The exhibition of the Famous African hermaphrodite

Advertisements of an African hermaphrodite appeared regularly in London newspapers throughout 1741.[9] For the price of two shillings and sixpence the 'Famous African' was to be seen at the Golden Cross, near Charing Cross, from nine in the morning until three in the afternoon, and from four in the afternoon until ten at night up some stairs in the corner of the Little Piazza, Covent Garden.[10] There were also exhibitions at private houses 'if any family desired so'. Advertisements portrayed the 'Famous African' as a curious combination of typically male and female features:

> Being a native of Angola, about 25 years old; of Masculine Features, which however seem perfectly Feminine in the Circumstances of Smiling or Joy. The voice, when deliver'd in a low tone, is quite a woman's; if aloud a Man's, and remarkably so in Expressions of Energy and Passion. The Chest and Shoulders are very robust and spread; the Paps hard and flat; the Muscles above the Elbow vastly strong and brawny; the Arms and Hands neat and slender; the Thighs and Legs a perfect Model of Female Proportion.[11]

The description of physical appearance in English was followed by a description of the reproductive parts in Latin, highlighting the African's double nature.[12] Usually, shows of human beings regarded as monsters were advertised in English and the use of Latin in this case was definitely related to the sexual nature of the 'monstrosity'. To a popular audience, Latin kept concealed what was normally hidden. At the same time, it drew attention to what might be revealed through live exhibition. Likewise, the advertisement's use of two languages differentiated the two kinds of audience potentially interested in the exhibition, a learned and curious one from the merely curious.

Considering the duration of the exhibition (according to the advertisements at least ten successive months), together with its long opening hours (about fourteen hours), the African hermaphrodite must have been a very popular show. This popularity is all the more remarkable if we bear in mind that it was an individual exhibition. Moreover, the African did not display any other supplementary gifts or perform any remarkable feats, like the majority of other human 'monsters' exhibited at the time.[13]

The reasons for the extraordinary popularity of the African hermaphrodite presumably had much to do with sexual voyeurism. Moreover, because it was a curiosity of nature, the sexual *frisson* could be veiled. The show's popularity seems also to have been linked to issues of race. In the different advertisements published, the noun 'hermaphrodite' was never mentioned. Instead, the expressions 'Famous African' and 'Native of Angola' were used and highlighted in capital letters. A black both male and female, made it possible to advertise the Famous African as a 'Mundi Mirabile Monstrum'.[14] The role of African origins in the exhibition was related to the diffusion, mainly through travel literature, of ideas about Africans having an excessive and unruly sexuality.[15] These were attributes closely associated with the monstrous. The African origins of this hermaphrodite reinforced, therefore, the strange and extraordinary nature of what was to be seen.

It is difficult to uncover details concerning the exhibition of the African hermaphrodite. It would be interesting to know, for example, if he/she was exhibited totally naked or not. Advertisements called attention to 'a particular description of the parts which distinguish the sexes' being given at the place of exhibition.[16] The show provided not only a panoramic view, but a guided tour into the recondite regions of the African's body. Later versions of the advertisements announced the selling of 'a curious engraved figure taken from, and as big as the Life, at 1 shilling each, to be sold to none but those who have seen or come to view the original'.[17]

According to the advertisements, the audience for the exhibition of the African hermaphrodite was not restricted to men. They also made explicit reference to the attendance, not only of 'Persons of the Highest Distinction' and of 'the most Learned', but also the 'Curious of both Sexes'.[18] The female audience was provided with special guidance by 'a prudent Person of their own Sex' on Mondays, Wednesdays and Fridays.[19] As prices were reduced in the later months of the exhibition a significant number of people from lower social backgrounds might well have attended the exhibition, too.

At the end of May and during the month of June, the Famous African was also advertised as 'The Phoenix', the bird of ancient Egyptian myth which had become a symbol of androgyny and infinite self-perpetuation.[20] The symbol of the phoenix was more likely to appeal to a learned, discerning audience. The positive mythical and artistic associations of androgyny could also have been used to mask the potentially monstrous bisexual nature of the African hermaphrodite.

Thus, throughout 1741, new elements were incorporated into the rhetoric used in the advertisements. Of these, one of the most noticeable is the reference to medical opinions concerning the nature of the 'Famous African'. In February 1741, soon after a new edition of William Cheselden's *Anatomy of the Human Body* was published, a note was added to the advertisements for the exhibition drawing attention to the inclusion of an engraving of the African (see Fig. 4):

William Cheselden, Esq., Surgeon to her late Majesty, having seen this subject, and caus'd an exact draught to be made of the Parts, has been pleas'd to give an engrav'd Figure thereof in the new Edition of his Anatomy (Pag. 314) where he pronounces it to be a wonderful mixture of both Sexes.[21]

Cheselden's *Anatomy* was a widely diffused medical work which went through thirteen editions between 1713 and 1792.[22] The original illustrations had been engraved by various artists but the plates of the fourth edition (1730) were all re-engraved by Gerard van der Gucht (1696–1776), an artist of Flemish origin who also worked for other eighteenth-century medical doctors.[23] According to the advertisements of the African, Van der Gucht was also the artist of an engraved figure of the African sold at the place of exhibition.

New versions of the advertisements which appeared at the end of March, added the views of other medical authorities to William Cheselden's:

James Douglas, M.D. and F.R.S., after strictly examining this subject, pronounces it FEMALE.
    John Freake, Surgeon to St. Bartholomew's Hospital, and F.R.S., after several times viewing and examining this subject, pronounces it MALE.[24]

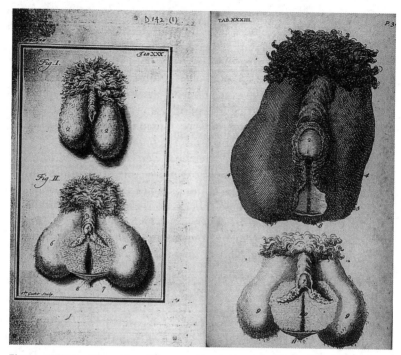

*Fig 4.* William Cheselden, 'Figures of hermaphrodites', from *The Anatomy of the Human Body*, engravings. *Left*: 'The parts of an Hermaphrodite, in which appeared as much of the mixture of the Sexes as could be', 2nd edition, 1722 (also in the 4th and the 5th editions). *Right*: 'The part of an Hermaphroditic negro, which was neither Sex, but a wonderful mixture of both', 6th edition, 1741. This illustration was published until the 13th and last edition in 1792.

Thus, what was to one medical testimony a 'wonderful mixture of both sexes' became no more than a 'female' to one and a 'male' to another.[25] The gaze of medical doctors and surgeons was not conducive to a similar understanding of sexual difference. Physical appearances could be interpreted in very different ways. In this distinct, public way, the 'Famous African' defied and problematized the role and authority of the medical profession in mediating sexual difference. Moreover, what was also at stake was the very possibility of the existence of beings with a double sexual nature.

James Parsons was one of the medical doctors who sought to intervene in the debate and give a definitive and clear answer to the question of the existence of human hermaphrodites. For three shillings and sixpence, almost the same price as the exhibition of the African, he published a treatise on hermaphrodites in July 1741 where he argued that they were all mere women with genital deformations. As he acknowledged, the work was mainly motivated by the exhibition of the 'Famous African' in London.[26] It was also a kind of 'Enlightenment guide' to what should be seen at Golden Cross. What was to be seen was no longer the 'Famous African' hermaphrodite, but just an 'Angolan woman'. In contrast to the advertisements of the exhibition, one of Parsons' strategies in the construction of this new identity was to erase any evidence of male features from the physical portrait of the 'Famous African':

> She is about six and twenty years old, has no Beard on her Chin, nor any Thing masculine in her Countenance; her Arms above the Elbow are thick and fleshy, as many Women are, but soft; her Breasts are small, her voice effeminate in the common Tone of speaking, and it was reported she has often been laid with Men.[27]

## Visual representations of human hermaphrodites

*A Mechanical and Critical Enquiry into the Nature of Hermaphrodites* (1741) was advertised in the *Daily Gazetteer* as a work with 'several cuts suitable to the subject, particularly one of the woman lately brought from Angola, exactly delineated from Life'.[28] It was also announced in the same month in *The Gentleman's Magazine* and the *Bibliothèque Raisonnée des Ouvrages des Savants de L'Europe*.[29] Although the work went through only one English edition, Parsons seems to have had in mind a wide audience for his work, and it seems to have been widely diffused.[30] He was careful to present all his quotations on hermaphrodites by ancient and early modern authors both in the original and in English. The choice of J. Walthoe, a publisher of didactic literature, was illustrative of Parsons' effort to distance his work from licentious and erotic literature in which hermaphrodites were sometimes mentioned.[31] More significant for the legitimation of Parsons' work, was its dedication to the President of the Royal Society, and to the Council and Fellows.[32] Indeed, a few months after its publication, Parsons presented and introduced his book at the Royal Society, which gave him an official acknowledgement.[33]

The illustrations in Parsons' work had a prominent role in his argumentative strategy to prove that all cases of hermaphrodites were in fact merely women with genital deformations. In his view, their visual immediacy and explicitness made them his most powerful weapon 'to overturn Errors, and undeceive the Crowd that is hurried along the through the Mazes and Labyrinths of Misrepresentations'.[34] As Parsons commented in his work, clear and particular anatomical representations were indispensable in giving credit and authenticity to reports of unusual occurrences of nature; only then could they be perceived as true mediators in the perception of nature:

> I cannot devise by what means Credit should be given to such Narrations as these, which so digress from human Nature's Laws, when not accompanied with a very nice and particular anatomick Description of such Parts.[35]

In Parsons' view, anatomical representations had not only to be based on first-hand observations but devoid of any subjectivity. The main illustration of his treatise is representative of his ideals of objectivity and realism. Significantly, it is entitled 'A View of the External Parts of Generation in the African woman, that was brought lately from Angola, exactly delineated from Life, and well engraven' (see Fig. 5). The huge dimensions of this illustration and its naturalistic mode of representation were important in its authentication. However, as the title indicates, the process of engraving was even more important in conveying detail and immediacy which contributed further to its sense of realism.[36]

The representations of the Famous African, like the eighteenth-century illustrations of hermaphrodites, focused only on the genitalia.[37] His/her physical appearance and sexual inclinations were hidden in a process which simplified sexual identification. The 'depersonalization' of the hermaphrodite was thus crucial for the construction and diffusion of medical ideas.

Another strategy used by Parsons to overcome the singularity of the Famous African was the juxtaposition of his/her left and right leg and two illustrations of another hermaphrodite exhibited at Charing Cross in the early eighteenth century. The inclusion of these two other figures in the illustration reinforced the 'depersonalization' of the Angolan and her reduction to another example of genital deformation.[38] This other case had been studied and described by James Douglas to whom Parsons acknowledged his debt. Douglas remarked, however, on the sexual preferences of this 'pretended hermaphrodite' towards women:

> Her inclinations of late are very much towards her own sex with whom she has been attempted to convene after the manner of the Tribades or fricatrices, but cannot entertain the thoughts of suffering a man.[39]

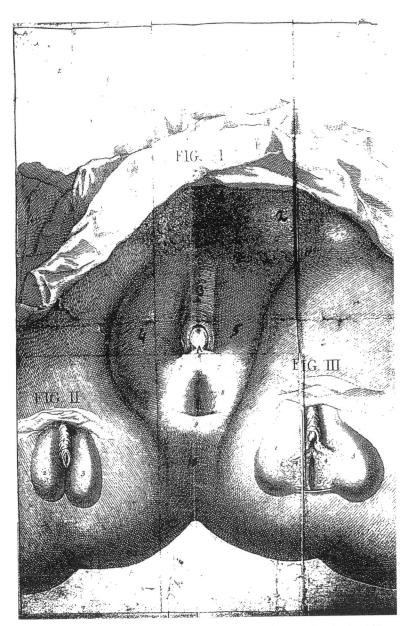

*Fig 5.*   James Parsons, 'A View of the External Parts of Generation in the African Woman that was brought lately from Angola', from *A Mechanical Enquiry into the Nature of Hermaphrodites*, 1741, engraving.

Douglas' illustrations of the hermaphrodite exhibited in London in the early eighteenth century were also used by William Cheselden in a completely different way. The engravings were published in four editions of Cheselden's *Anatomy of the Human Body* (from 1722 to 1741) (see Fig. 4). Douglas' opinion was also acknowledged by Cheselden but he pronounced the two cases to be examples of a perfect mixture of the two sexes.[40]

The other illustration included in Parsons' work represented the Famous African with the labia shut or closed. This illustration was included with the purpose of showing 'how easily the ignorant and superstitious might be deceived at the sight of such Parts, when in the same Circumstances with this Subject, and the *Labia Pudendorum* not separated'.[41] It was, thus, used as a warning and a reminder of the usefulness of the anatomist in the exposure of 'truth' to 'the vulgar'. Only a qualified anatomist would be able to go beyond appearances and have access to the true order of nature.

The illustrations of the Famous African intended to illustrate that genital deformation was not only due to the extraordinary dimension of the clitoris, but also to the unusual appearance of the right labium. In Parsons' and James Douglas' understanding, this appearance had a tumoral origin but was explained in a different way by other physicians such as Hans Sloane.[42] It was these two extraordinary characteristics which made it so crucial to represent the orifice of the vagina in the larger illustration, as well as representing the clitoris in a way that showed that it was not perforated. In this way, the representation was 'proof' that the 'Famous African' had nothing 'but what is common to every Woman'.[43] Nevertheless, Parsons recognized that only 'a fair dissection of such a subject … [and by] the best anatomists' would provide a conclusive understanding of the extraordinary appearance of the right labium.[44]

It was primarily this unusual formation of the right labium that rendered the Famous African such an ambiguous and controversial case in the identification of sex.[45] This made Parsons' observations about the similarities of the genitals of female foetuses to hermaphrodites crucial in his argument about the non-existence of human hermaphrodites. Unlike the case of the 'Famous African', the resemblance of female foetuses to hermaphrodites was, to Parsons, an observed law of nature.[46] At a meeting of the Royal Society held a few months before publication of his book, he had exhibited some female foetuses in which the size of the clitoris resembled those of the hermaphrodites closely.[47] According to the minutes of this meeting, it was this observation particularly that led Parsons to conclude that 'the hermaphrodite state is nothing else but a preternatural proportion of the female parts, owing to same accidental cause, such as the heat of the climate or the like which hinders the natural contraction of those parts, and dispenses them to an unnatural growth'.[48] Nevertheless, it was the sensational and much debated case of the 'African hermaphrodite' that occupied the more prominent place in his book on human hermaphrodites.

## The medicalization of human hermaphrodites

Parsons' understanding of human hermaphrodites is an illustrative case of what Michel Foucault described as the persistent and stubborn attempt of modern Western societies to decipher 'a true sex'.[49] The view that certain of the so-called hermaphrodites were in fact women with deformed genitals had been the argument of a number of seventeenth-century anatomists, most famously Jean Riolan, the younger.[50] Yet, for Parsons and others who shared his interpretation, this happened not only for specific occurrences but was a general law of nature. Another difference was that in the eighteenth century anatomists increasingly asserted their prerogatives over midwives, surgeons and physicians in negotiating the sex to which the 'hermaphrodite' belonged.[51]

In accordance with Enlightenment ideals, Parsons' main concern, as stated in the preface of his book, was the repudiation of superstition and vulgar errors:

> I have this for my Plea, that the expulsion of superstitious Mysteries and Errors, occult Causes, and, in fine, the Promotion of Truth, in some Parts of Natural Knowledge, to the utmost of my Power, are my sole Intention.[52]

The introduction provided, however, an extensive historical survey of laws relating to hermaphrodites.[53] The importance of this legal aspect in Parsons' work is indicative of how his medical understanding of hermaphrodites was closely connected with social concerns and anxieties about 'gender' identification. Indeed, to Parsons, the extraordinary size of the clitoris was not only the cause of false notions, it was also the source of two other evils: 'the hindering of the Coitus, and women abuse of them with each other'.[54] Therefore, he agreed how wise it was 'to cut or burn them off while Girls are young, and at the same time never entertain the least Notion of the Existence of any other nature besides the Female in those subjects who are thus deprived of that useless Part'.[55] The performance of the excision of the clitoris had been mentioned occasionally by medical authors such as Ambroise Paré.[56] However, the reduction of all human hermaphrodites to women with genital deformations contributed to its legitimation. Surgeons were thus supposed to have a prominent role not only in eradicating monstrous appearances but also 'monstrous' sexual practices between women. These opportunities were seized upon by surgeons such as George Arnaud in his *A Dissertation on hermaphrodites* (1750):

> if nature departs from her usual source, and forgets herself, she may, by the aid of this art, be redressed, and new ways, capable of putting her in the state of perfection she ought to have, be paved out for her.[57]

According to his work, Arnaud did in fact perform a surgical operation on a thirty-five-year-old hermaphrodite.[58] His patient's genitals had the appearance of a vagina but this did not have any orifice. Nevertheless, he/she had regular discharges of menstrual blood which were 'obliged to pass every month through the anus' thus causing the hermaphrodite pain and a general 'bad state of health'.[59] Arnaud's operation consisted in opening an orifice in what seemed to be the vagina of the hermaphrodite but was only partially successful. After some months, his patient started suffering once more from the same unusual symptoms.

The use of surgical intervention in the medicalization of human 'hermaphrodites' was similar to the treatment for other monstrosities such as the removal of monstrous horns. At one of the meetings of the Royal Society of London, the surgeon John Stevens showed several 'tumours' that he had taken off the head of a maid living at Chelsea.[60] He also suggested that the horn-like substances on the head of a woman who was exhibited in London were similar in kind.

One important consequence of the reduction of monstrous formations to pathological instances, was that this contributed to a separation of social spheres in the experience of the monstrous. The notion of monsters as singularities of nature presented in public exhibitions became increasingly part of a culture of entertainment associated with the unlearned and 'the vulgar'. In this sense, the medicalization of hermaphrodites illustrates a general trend in the understanding of monstrosity which was to become more apparent in the second half of the eighteenth century. Throughout this period, other monstrous occurrences were understood to be pathological conditions of the body. The inclusion of some of these cases in Mathew Baillie's *Morbid Anatomy* (1793) is indicative of this process.[61]

In accordance with an Enlightenment sensibility, the refutation of human hermaphrodites was associated with a quest for wider medical authority in the eradication of superstition and false beliefs related to monstrous and singular productions of nature. Nevertheless, the case of the 'Famous African' shows that medical understanding of human hermaphrodites was by no means obvious or uncontroversial. Indeed, as pointed out before, some medical doctors and surgeons continued to believe in their existence or to explain their existence in conflicting ways. Moreover, various editions of Cheselden's *Anatomy of the Human Body* continued to reiterate this view throughout the eighteenth century. Parsons' treatise on hermaphrodites did not end the disputes about their existence, but the traces of these other medical accounts of sexual identity become much weaker. Robert James' short reference to hermaphrodites in the first edition of his *Medical Dictionary* (1743–5) is illustrative of the general tone of the medical literature in this regard:

> As I look upon all the histories related to hermaphrodites to be merely imaginary, I shall only observe, that of many I have seen who have been reputed to be so, I have met with none who were more than mere women, whose clitoris was grown to an exorbitant size, and whose *Labia Pudendorum* were preternaturally tumid.

In a similar way, the entry on 'hermaphrodites' in the second edition of James Fytler's editon of the *Encyclopaedia Britannica* (1781) was enlarged with a reference to the publication of James Parsons' work on hermaphrodites:

> Dr. Parsons has given us a treatise on this subject, where he endeavours to shew the notion of hermaphrodites to be vulgar error; and, in particular, the *Anglo-Negro*, shewn about this town some years ago, was a woman whose clitoris was overgrown.[62]

The 'Famous African', half-naturalized as 'Anglo-Negro' in Fytler's words, had now become a paradigmatic example of the false hermaphrodite.

Yet by this time John Hunter had already published an article in the *Philosophical Transactions* which presented a much less radical view on the subject.[63] For Hunter, human hermaphrodites existed but in a reduced number in comparison with other species.[64] Moreover, he did not confine himself to the study of the generative differences of hermaphrodites. He was also interested in the general problem of the development of secondary sexual characteristics in animals. In his 'Account of an Extraordinary Pheasant', he remarked that there is often a change in the secondary properties from one sex to another, the female now and then assuming the peculiarities of the male with respect to the secondary properties.[65] He pointed out that the extraordinary pheasant, which had belonged to a lady well known to the President of the Royal Society, was a perfect female with respect to the generative parts, but assumed a male appearance at a certain time. Hunter's conclusion for these kind of occurrences was that there is a disposition in the female to gradually approximate the male, at least in the secondary properties. The case gave him a more controversial suggestion: 'that female character contains more truly the specic properties of the animal than the male; but the true character of every animal is that which is in both sexes, *viz.* a natural hermaphrodite, or an animal or neither sex, *viz.* a castrated male or spayed female'.[66]

In the nineteenth century, the figure of the hermaphrodite continued to challenge medical notions of sexual difference and gender identification. Alice Dreger has shown that by the end of this period, around 1870, a consensus began to solidify around the notion that the single reliable marker of 'true sex' in doubtful cases was the gonad, that is, the ovary and testicle.[67] Such a definition was generally used by medical men until 1915 and one of its consequences was that it virtually eliminated 'true' hermaphroditism. Moreover, surgical operations on persons of doubtful sex were widespread in the nineteenth century, something which effectively contributed to erase the figure of the hermaphrodite from medical records and society.

## Conclusion

The reduction of human hermaphrodites to women with genital deformations presented in Parsons' work and shared by many of his contemporaries fitted the Aristotelian notion of the female body as a 'pathological' deviation from the male.[68] As Ludmilla Jordanova has remarked, 'the concept of woman carried the burden of difference'.[69] Moreover, it fulfilled an important social role. The refutation of human hermaphrodites enabled the common order of nature to be preserved and encouraged the eradication of sexual practices traditionally associated with hermaphrodites. If supposed hermaphrodites were no longer monsters, practices like sodomy and masturbation were now even more radically monstrous.[70] They were threatening to the social order in challenging male privileges and the subordinate status of women in society.

The reduction of hermaphroditism to a mere pathological state of the body also contributed towards redefining the importance of moral monstrosity and its relationship to the physical domain. Implicitly or explicitly, the association between physical and moral monstrosity was often present in relations of monstrous births published in popular broadsheets.[71] Some of these accounts described monstrous occurrences as a punishment for the illicit behaviour of the mother.[72] They were used as an *exemplum* in the service of a moral and religious lesson.[73] In a similar way, eighteenth-century medical and paramedical anti-masturbatory literature presented cases of women with extraordinarily sized clitorises as the result of their 'self-pollution' and moral depravity.[74] Indeed, in his *Onanism* (1760), Simon André Tissot described a hermaphrodite as being 'by all look'd upon as a Creature of Vile Deformity'.[75] In such literature, personal histories were also used as an example of moral depravity.[76]

I have also noted that the denial of human hermaphrodites and their reduction to women with genital deformations presupposed the anatomist's privileged access to 'the true identity' of their nature. In this way, human 'hermaphrodites' were used by some anatomists to strengthen their authority in the eradication of superstition and false beliefs associated with monstrous occurrences. At the same time, the social implications of the 'medicalization' of human hermaphrodites were framed within an appeal to the moral authority of nature.[77] Therefore, this case especially illustrates the use of the discursive arsenal of enlightened rationality to contain and eradicate the monstrous from English culture and society. I have shown, however, that the denial of human hermaphrodites was not uncontested or uncontroversial within the medical profession and that they continued to challenge an Enlightenment sensibility based on the search for order and regularities throughout the eighteenth century.[78]

## Notes

1. On hermaphrodites in the sixteenth and seventeenth centuries, see L. Daston and K. Park, 'Hermaphrodites in Renaissance France', *Critical Matrix*, 1 (1985),

1-19, and L. Daston and K. Park, 'The Hermaphrodite and the Orders of Nature: Sexual Ambiguity in Early Modern France', *GLQ: a Journal of Lesbian & Gay Studies*, 1 (1995), 419–38. See also K. P. Long, 'Hermaphrodites Newly Discovered: the Cultural Monsters of Sixteenth-Century France', in J. Cohen (ed.), *Monster Theory: Reading Culture* (Minneapolis and London: University of Minnesota Press, 1996), pp. 1183–201.

2. The principal exceptions were serving as a lawyer, a judge, or a rector of a university. See J. Parsons, *A Mechanical and Critical Enquiry into the Nature of Hermaphrodites* (London, 1741), pp. xxxix–xl.

3. R. Trumbach, 'London's Sapphists: from Three Sexes to Four Genders in the Making of Modern Culture', in G. Herdt (ed.), *Third Sex, Third Gender: Beyond Sexual Dimorphism in Culture and History* (Zone Books, 1994), pp. 112–41.

4. Daston and Park, 'The Hermaphrodite and the Orders of Nature', pp. 420–4.

5. On Hippocratc and Aristotelian conception theories, see M. Boylan, 'The Galenic and Hippocratic Challenges to Aristotle's Conception Theory', *Journal of the History of Biology*, 17 (1984), 83–112.

6. In his *Generation of Animals,* trans. A. L. Peck (Cambridge, Mass. and London: Harvard University Press, 1990), pp. 427–8, Aristotle discusses hermaphrodites in the context of multiple births.

7. See A. Paré, *Des Monstres et Prodiges* (1573), J. Céard (ed.) (Genève: Librarie Droz, 1971), pp. 24–8; J. Duval, *Des hermaphrodites* (Rouen, 1682); J. Riolan, *Discours sur les hermaphrodites* (Paris, 1614); K. Bauhin, *De hermaphroditorum monstrosorumque partum natura* (Oppenheim, 1614); and N. Venette, *La Génération de l'homme ou Tableau de l'amour conjugal considéré dans l' état du marriage* (Paris, 1687) (*The Mysteries of Conjugal Love Reveal'd* (London, 1707)), pp. 453–70.

8. I am here specially referring to the eighteenth-century growing competition between midwives and man-midwives and between surgeons and physicians. See A. Wilson, *The Making of Man-midwifery: Childbirth in England, 1660–1770* (London: UCL Press, 1995); and S. Lawrence, 'Anatomy and Address: Creating Medical Gentlemen in Eighteenth-Century London', in V. Nutton and R. Porter (eds), *The History of Medical Education in Britain* (Amsterdam: Rodopi, 1995), pp. 199–228.

9. *London Daily Post and General Advertiser* from January to August 1741. The *Daily Gazetteer* was also searched for this period, but no advertisement of the African hermaphrodite was published in this newspaper during 1741.

10. In the last months of exhibition the price was reduced to one shilling.

11. *London Daily Post and General Advertiser*, 15 January 1741.

12. 'Scroti quippe vice fuguntur Vulvae Labia, Testiculos binos largos & mobiles involventia. Clitoridis loco Penis viribilis ad quatuor minimum prominet digitos, ubi (quod saepius sit) erigitur; imperforatus quidem, sed ex corporibus cavernosis, Glande, Praeputio & Fraemulo Vere conflatus. Adsunt praeterea, planè ac in Mulieribus, Meatus urinarius, & vagina satis ampla; Hujusque tamen nulla huic ANDROGYNAE effluxerunt catamenia', ibid.

13. See R. Altick, *The Shows of London* (London: Belknap Press of Harvard University Press, 1978), pp. 34–49.

14. See L. Schiebinger, 'The Anatomy of Difference: Race and Sex in Eighteenth-Century Science', *Eighteenth-Century Studies*, 23 (1990), 387–405.

15. L. Africanus, *Historie of Africa* (London, 1526), widely available in Latin in England and translated in 1600 by J. Pory, was one of the influential works which contributed to the diffusion of these views.

16. *London Daily Post and General Advertiser*, 24 March 1741.

17. Ibid., 2 May 1741.

18. Ibid., 15 January 1741.

19. This note is added in the advertisement after 10 March 1741. After being exhibited in London, 'The Famous African' went on display during York's race week. On eighteenth-century public exhibitions of this kind in the provinces, see J. J. Looney, 'Cultural Life in the Provinces: Leeds and York, 1720–1820', in A. L. Beier, D. Cannadine and J. M. Rosenheim (eds), *The First Modern Society: Essays in English History in Honour of Laurence Stone* (Cambridge: Cambridge University Press, 1989), pp. 483–510.

20. On mythical monsters, see J. Cherry (ed.), *Mythical Beasts* (London: British Museum Press, 1995). On hermaphrodites in ancient Greece, see M. Delcourt, *Hermaphrodite, mythes et rites de la bisexualité dans l'antiquité classique* (Paris: Presses Universitaires de France, 1958).

21. *London Daily Post and General Advertiser*, 6 February 1741.

22. Cheselden made some alterations and additions in the first few editions but after the sixth edition, the text and illustrations remained constant.

23. K. B. Roberts and J. D. W. Tomlinson, *The Fabric of the Body: European Traditions of Anatomical Illustration* (Oxford: Clarendon, 1992), pp. 412 and 433.

24. *London Daily Post and General Advertiser*, 24 March 1741. Freake argued against Parsons' view on hermaphrodites at two meetings of the Royal Society. See *Royal Society's Journal Books,* 24 February 1743; 3 March 1743.

25. Throughout the eighteenth century, William Cheselden's view was reiterated in the various editions of his successful *Anatomy of the Human Body*.

26. Parsons, *Enquiry*, p. liv.

27. Ibid., pp. 134–5.

28. *Daily Gazetteer*, 3 July 1741.

29. *The Gentleman's Magazine*, July, 1741; and *Bibliothèque Raisonnée des Ouvrages des Savants de L'Europe*, pour les Mois de Julliet, Août & Septembre, Tome 27, Amsterdam 1741, p. 220.

30. A letter written by an unnamed surgeon concerning Parsons' work on hermaphrodites was published in the *Gentlemen's Magazine* of February, 1744, p. 91. As I will show later, the work was also often mentioned in medical dictionaries and encyclopaedias of the period. On eighteenth-century readership of natural philosophical and natural historical books, see G. S. Rousseau, 'Science Books and their Readers in the Eighteenth Century', in I. Rivers (ed.), *Books and their Readers in the Eighteenth Century* (Leicester: Leicester University Press and St. Martin's Press, 1982), pp. 197–256.

31. Compare, for example, with *A Treatise of the Use of Flogging in Venereal Affairs ... to which is Added a Treatise of Hermaphrodites* published by Edmund Curll in 1718. On eighteenth-century medical and paramedical erotic literature, see P. Wagner, 'The Discourse on Sex – or Sex as Discourse: Eighteenth-Century Medical and Paramedical Erotica', in G. S. Rousseau & R. Porter (eds), *Sexual Underworlds of the Enlightenment* (Chapel Hill: University of North Carolina Press, 1988), pp. 46–68 and G. S. Rousseau and

R. Porter, *Eros Revived: Erotica of the Enlightenment in England and America* (London: Secker & Warburg, 1988), especially pp. 1–47.

32. See Add. MS 4437, British Library. This letter was also published in the *Philosophical Transactions* with slight modifications. See James Parsons, 'A Letter Giving a Short Account of *a Mechanical Critical Inquiry into the Nature of Hermaphrodites*', *Philosophical Transactions*, 16 (1739–40), pp. 650–2.

33.   *Royal Society's Journal Books*, 29 October 1741.

34. Parsons, *Enquiry*, p. xv.

35. Ibid., p. 90.

36. As William Ivins has remarked, 'The lines used in the process of engraving could be very fine and very close together, as compared to those on any wood-blook, and still yield a sufficient quantity of clear impressions on the paper available'. W. Ivins, *Prints and Visual Communications* (Cambridge, MA & London: MIT, 1989), p. 165. On the role of visual representations in the validation of extraordinary phenomena of nature, see P. Fontes da Costa, 'The Making of Extraordinary Facts: Authentication of Singularities of Nature at the Royal Society of London in the Eighteenth Century', *Studies in the History and Philosophy of Science*, 33 (2002), 265–88.

37. On the use of the fragmented female body in anatomical illustrations, see L. Jordanova, 'Gender, Generation and Science: William Hunter's Obstetrical Atlas', in *Nature Displayed: Gender, Science and Medicine, 1760–1820* (London and New York: Longman, 1999), pp. 183–202; and D. Petherbridge and L. Jordanova, *The Quick and the Dead: Artists and Anatomy* (London: National Touring Exhibitions, 1998), pp. 84–6.

38. Lorraine Daston and Peter Galison have recently shown the widespread use, in the eighteenth century and beyond, of representations of individual cases as exemplary and illustrative of broader classes of phenomena, which they label 'characteristic images'. See L. Daston and P. Galison, 'The Image of Objectivity', *Representations*, 15 (1992), 81–128, especially, pp. 93–8.

39. DF. 60 (2), Hunterian Collection, Glasgow University Library.

40. W. Cheselden, *The Anatomy of the Human Body* (London, 1722 and 1741), p. 319.

41. Parsons, *Enquiry*, p. 155.

42. 'As to the said Tumour in the Labium, several of the Learned are divided about it, and their different Opinions amount to three, viz. 1. That such are Testes like those in Men. 2. That they are Herniae of the Ovaria. 3. That they are Glands of an indolent Nature, void of any Use, fallen from the Groins, and grown inordinately large and hard from the same Cause that enlarges any other neighbouring Parts that exceed their natural size. To the first of these Mr. Cheselden, and, I am told, some others in Town, seem to assent. The second is the Opinion of Dr. Douglas and the last is the Conjecture of Sir Hans Sloane', ibid., pp. 142–3.

43. Ibid., p. 141.

44. Ibid., pp. 142–3.

45. As Ludmilla Jordanova has remarked, 'It may be that the sex/gender distinction has little relevance for the eighteenth century, although certainly there were attempts to differentiate between inherent and acquired characters. But the idea that there were two kinds of difference (gender, sex) between men and women was foreign to them'. Jordanova, 'Sex and Gender', in

C. Fox, R. Porter and R. Wolker (eds), *Inventing Human Science: Eighteenth-century Domains* (Berkeley: University of California Press, 1995), p. 156. See also L. Jordanova, 'Gender and the Historiography of Science', *British Journal for the History of Science*, 26 (1993), 469–83; and D. Lupton, *Medicine as Culture: Illness, Disease and the Body in Western Societies* (London: Sage Publications, 1994), pp. 24–30.

46. Parsons, *Enquiry*, p. 147.

47. Parsons' observations on female foetuses draw attention to the increasing role of the study of foetal development in the understanding of monsters in the eighteenth century. The dimension of time complemented the spatial analysis provided by the dissection of monstrous bodies.

48. *Royal Society's Journal Books*, 30 April 1741.

49. M. Foucault, 'Introduction', in *Herculine Barbin: Being the recently discovered memoirs of a nineteenth-century french hermaphrodite* (1978), trans. Richard McDougall (Brighton: Harvester Press, 1980). See also M. Foucault, *The History of Sexuality* (1976), trans. R. Hurley (London: Penguin Books, 1990), especially vol. I, pp. 3–13.

50. J. Riolan the younger, *Discours sur les hermaphrodites* (Paris, 1614).

51. For a comparison with the previous period, see K. Park, 'The Rediscovery of the Clitoris: French Medicine and the Tribade, 1570–1620', in D. Hillman and C. Mazzio (eds), *The Body in Parts: Fantasies of Corporeality in Early Modern Europe* (New York and London: Routledge, 1997), pp. 171–93; L. Daston and K. Park, 'Hermaphrodites in Renaissance France', and 'The Hermaphrodite'.

52. Parsons, *Enquiry*, p. iv. On the ethos of the Enlightenment, see L. Daston, 'Afterword: the Ethos of the Enlightenment', in W. Clark, J. Golinski, and S. Schaffer (eds), *The Sciences in Enlightened Europe* (Chicago and London: University of Chicago Press, 1999), pp. 495–504.

53. Legal questions discussed included 'whether a Man's or Woman's Name should be given to an Hermaphrodite at its Baptism?'; 'How often should an hermaphrodite confess?'; 'Can an Hermaphrodite contract marriage?'; 'Can an hermaphrodite be a witness?'; 'Can hermaphrodites be promoted to holy Orders?'; 'Can an hermaphrodite be Rector of a University?'; 'Can an hermaphrodite be a Judge?' Parsons' book also included three chapters and a conclusion. The first chapter presented his arguments against the existence of hermaphrodites in human nature. The second presented a critical account of the causes that several authors used in the explanation of the origin of hermaphrodites. Finally, the third chapter consisted of a critical view of the histories of hermaphrodites given by several authors.

54. Parsons, *Enquiry*, p. 10.

55. Ibid., p. 11.

56. Paré, *Des Monstres et Prodiges*, p. 27. On the excision of the clitoris as 'corrective surgery', see T. Laqueur, 'Amor Veneris, vel Dulcedo Appeletur', in M. Feher, R. Naddaff and N. Tazi (eds), *Fragments for a History of the Body* (New York: Zone, 1990), pp. 113–20; and Park, 'The Rediscovery of the Clitoris', pp. 183–4.

57. G. Arnaud, *A Dissertation on Hermaphrodites* (London, 1750), p. 466. Arnaud performed a surgical operation on the genitals of an 'hermaphrodite' patient of his. See ibid., pp. 463–5.

58. Ibid., pp. 463–5.

59. Ibid., p. 458 and p. 460.

60. *Royal Society's Journal Books*, 13 May 1736. See also, G. Ash, 'A Letter Concerning a Girl in Ireland with Horny Excrescences', *Philosophical Transactions of the Royal Society of London*, 15 (1685), 1202–4; and E. Home, 'Observations on Certain Horny Excrescences of the Human Body', *Philosophical Transactions of the Royal Society of London*, 81 (1791), 95–105.

61. Mathew Baillie includes in his work cases of 'the hymen imperforated', 'the clitoris enlarged', and 'the external labia growing together'. See M. Baillie, *The Morbid Anatomy of Some of the most Important Parts of the Human Body* (London, 1793), ch. 22.

62. J. Fytler (ed.), *Encyclopaedia Britannica*, 2nd edn (London, 1781), vol. V, pp. 3621–3.

63. J. Hunter, 'Account of the Free Martin', *Philosophical Transactions of the Royal Society of London*, 69 (1779), 279–93.

64. In Hunter's view the number of hermaphrodites in a species was directly proportional to its place in the scale of being. Ibid., pp. 280–1.

65. J. Hunter, 'Account of an Extraordinary Pheasant', *Philosophical Transactions of the Royal Society of London*, 70 (1780), 527–35.

66. Ibid., pp. 530–1

67. A. Dreger, *Hermaphrodites and the Medical Invention of Sex* (Cambridge Mass, and London: Harvard University Press, 1998).

68. 'Anyone who does not take after his parents is really in a way a monstrosity, since in these cases Nature has in a way strayed from the generic type. The first beginning of this deviation is when a female is formed instead of a male, though this indeed is a necessity required by Nature, since the race of creatures which are separated into male and female has got to be kept in being'. Aristotle, *Generation of Animals*, trans. A. L. Peck (Cambridge, Mass. and London: Harvard University Press, 1990), p. 401. On the notion of the female body as 'pathologized', see Sander L. Gilman, *Sexuality: an Illustrated History* (New York/Chichester: Wiley, 1989), especially ch. 6. See also T. Laqueur, *Making Sex: Body and Gender from the Greeks to Freud* (Cambridge, MA: Harvard University Press, 1990), especially ch. 5. On representations of women in the Enlightenment, see S. Tomaselli, 'The Enlightenment Debate on Women', *History Workshop Journal*, 20 (1985), 101–24. See Also L. K. Friedli, 'Crossing Gender Boundaries in Eighteenth-Century England', unpublished PhD Dissertation (University of Essex, 1987), pp. 168–261.

69. Jordanova, 'Sex and Gender', p. 159.

70. On sodomy in the eighteenth century, see R. Trumbach, 'London's Sodomites: Homosexual Behaviour and Western Culture in the Eighteenth Century', *Journal of Social History*, 11 (1977), 1–33; and idem, 'London's Sapphists'. On tribadism, see E. Donoghue, *Passions Between Women: British Lesbian Culture, 1668–1801* (London: Scarlet Press, 1993), especially Ch. 1; and V. Traub, 'The Psychomorphology of the Clitoris', *GLQ: A Journal of Lesbian & Gay Studies*, 2 (1995), 81–113. For primary sources, see Ian McCormick (ed.), *Secret Sexualites: a Sourcebook of 17th and 18th Century Writing* (London and New York: Routledge, 1997).

71. In his essay 'On Deformity', Francis Bacon had explicitly described physical monstrosity as a cause of moral monstrosity, F. Bacon, *The Essays*, J. Pitcher (ed.) (Harmondsworth: Penguin, 1985), p. 191.

72. For an apology of physical deformity, see W. Hay, *Deformity: an Essay* (Dublin, 1754). Hay was himself physically deformed. For a comprehensive bibliography on the subject, see U. Bolte, *Deformität als Metapher: ihre Bedeutung und Rezeption im England des 18. Jahrunderts* (Frankfurt am Main: Lang, 1993), pp. 205–43. On the relationship between physical and moral deformity in English literature, see C. Baldick, *In Frankenstein's Shadow: Myth, Monstrosity, and Nineteenth-century Writing* (Oxford: Clarendon, 1987), pp. 1–10; and D. Todd, *Imagining Monsters: Miscreations of the Self in Eighteenth-century England* (Chicago and London: Chicago University Press, 1995), especially ch. 7.

72. See, for example, *The Miracle of Miracles or the Birth of a Strange Monster in Nature as ever Was Heard of it* (London (?), 1715 (?)).

73. See R. Chartier, 'The Hanged Woman Miraculously Saved: An Occasionel', in R. Chartier (ed.), *The Culture of Print: Power and the Uses of Print in Early Modern Europe*, trans. L. G. Cochrane (Cambridge: Polity Press, 1989), pp. 59–91.

74. S. A. D. Tissot, *Onanism; or, a Treatise upon the Disorders produced by Masturbation* (1760), trans. A. Hume (London, 1766), p. 163. J. B. Bieville presents a similar view in his *Nymphomania or a Dissertation concerning the Furor Uterinus* (*La nymphomanie: ou, Traité de la fureur utérine*, 1771), trans. E. S. Wilmont (London, 1775). On nymphomania, see G. S. Rousseau, 'The Invention of Nymphomania', in his *Perilous Enlightenment: Pre- and Postmodern Discourses, Sexual, Historical* (Manchester and New York: Manchester University Press, 1991), pp. 44–64.

75. Tissot, *Onanism*, p. 165.

76. See, in particular, the cases published in *A Supplement to the Onania; or the Heinous Sin of Self-Pollution, and all its Frightful Consequences*, 5th edn (London, 1729).

77. On the authority of nature in the Enlightenment, see A. E. Pilkington, 'Nature as Ethical Norm in the Enlightenment', in L. Jordanova (ed.), *Languages of Nature: Critical Essays on Science and Literature* (London: Free Association Books, 1986), pp. 51–85; and L. Daston, 'Afterwod: the Ethos of the Enlightenment', in W. Clark, J. Golinski, and S. Schaffer (eds), *The Sciences in Enlightened Europe* (Chicago and London: University of Chicago Press, 1999), pp. 503–4.

78. On the challenge of monsters and the order of nature in eighteenth-century England, see Fontes da Costa, 'The Experience of the Singular at the Royal Society of London, 1695–1752', unpublished PhD thesis (University of Cambridge, 2000), ch. 5.

# 9

# Jules Guérin Makes his Market: the Social Economy of Orthopaedic Medicine in Paris, c. 1825–1845

*Constance Malpas*

In 1826, Georges Duval, the creator of countless vaudeville triumphs, brought another smash hit to the Paris stage: a play entitled *Le tailleur des bossus, ou l'orthopédie; contrefaçon en un acte.*[1] It is difficult to do justice to this title in English. A literal translation might be 'The Hunchbacks' Tailor, or Orthopaedia: a Fraud in one act' – but this fails to convey the droll *double-entendre* that gave the work its satirical edge and comic appeal. The title is, in fact, a play on words: the orthopaedic 'tailor' is really an 'operator', a charlatan whose efforts to 'refashion' reality are good for nothing but a laugh.

Of course, Duval himself expected to get something more than a laugh out of the *Tailleur des bossus*. He expected royalties, and lots of them – that was the beauty of vaudeville theatre: popular success meant great profits. Duval's productions were calculated to appeal to a broad audience, to maximize their profitability, and that's exactly what makes them so interesting to us now: they tell us something about the preoccupations of a broad cross-segment of the urban population. By the mid-1820s, it seems, the correction of skeletal deviations and defects was one of them – not, I think, because many theatregoers felt themselves concerned as patients, or practitioners, but because orthopaedic medicine itself had come to stand for something more than a set of practices, or principles; it had come to represent the social and political economy of the marketplace, the commercial culture of an emerging bourgeois elite.

This is what I would like to explore here: the commodification of orthopaedic medicine – the marketing of a discipline devoted to recti-fying (quite literally, straightening) the structural defects of the body. I am interested here in the connection between the emergence of new medical markets, and the institutionalization of new medical special-

148

ties.[2] The remarks I have to offer are based on a simple, even obvious, observation, and that is that the disciplinary development of orthopaedic medicine in France was driven by the marketplace.[3] Between 1825 and 1845, countless new corrective technologies and therapeutic regimes were introduced – not just in France, but through-out Western Europe – for the treatment of locomotor defects and disabilities; enterprising physicians, as well as unlicensed practitioners, discovered a lucrative new niche market in a middle-class population eager to improve its social and physical 'mobility'.

I would contend that changes in the social, political and economic landscape of the nineteenth and twentieth centuries were more influential in shaping the development of orthopaedic medicine than any sudden increase in the population of military invalids, or *mutilés de guerre*. The primary focus of research and of technological develop-ments in the field of orthopaedic medicine, when the speciality first emerged as a viable profession for licensed practitioners in the early years of the nineteenth century, was not the rehabilitation of maimed veterans, but the correction of relatively minor defects like club-foot, or curvature of the spine. The success of the therapeutic techniques that were introduced in this period – and the standing of the discipline as a whole – depended upon demographic changes that accompanied the rise of the bourgeoisie in the eighteenth and nineteenth centuries, and brought a new population of patients to the doors of private *maisons de santé*. Academic notoriety and economic success were won by physicians who treated scoliosis and wry-neck, not those who tended to the wards of ill and undernourished children in the charity hospital. Had the techniques of extension, exercise and tenotomy been widely used in the public hospital, where patients were more likely to suffer from diseases like rickets or tuberculosis (both associated with severe skeletal deformities), the probable outcomes – fused or fractured vertebrae, irreversible paralysis – would surely not have contributed to the success of the discipline's institutionalization.

In fact, the techniques that brought fame and fortune to physicians like Jules Guérin, Vincent Duval and Henri-Sauveur Bouvier, were developed outside of the ideological (and architectural) confines of the 'clinic'. The *tailleur des bossus* did not operate in the surgical amphithe-atres of the Faculty of Medicine, but upon a broader stage – that of social drama and the *comédie humaine*. For this very reason, the successes (as well as the failures) of orthopaedic medicine were a matter of popular interest: concern about the *tailleur*'s ability to 'redress' social inequalities by restoring mobility and self-sufficiency reflected more general anxieties about the importance of both moral and physical fitness in defining the economic and cultural elite. In the post-Revolutionary and Napoleonic period, the correction of non-life-threatening defects became profitable in a way it had not been before. The 'medicalization' of deformity and disability, in other words, was a *commercial phenomenon* that involved exchange and negotiation between two, if not three, interested parties: patients, practitioners and the professional elite. In the early decades of the nineteenth century, exchanges between these three communities became more interesting (in an economic sense) than ever. Let us take a moment to consider why.

The Restoration, and especially the July Monarchy, were marked by the emergence of a new social elite: the bourgeoisie – a cash-rich urban population anxious to consolidate its social and political authority, and determined to establish its respectability.[4] This was a population concerned with the preservation and valorization of moral virtues that they considered the hallmark of their kind. It was also a population interested in conforming to the prevailing aesthetic norms, which favoured long necks, straight backs and shapely feet – all features that were set off by contemporary fashions.[5] I do not mean to suggest that aesthetic criteria were unimportant in 'fashioning' social elites before the period of the July Monarchy, but I do want to draw attention to the fact that norms of physical desirability shaped the 'social existence' of young men and women in the 1830s and 1840s in a way that even contemporaries found noteworthy. In the summer of 1835, for example, a number of Paris newspapers reported on the recent suicide of a young man who had put an end to his life because he foresaw that his physical deformity would prevent him from ever enjoying the fruits of social success.[6]

Even a relatively minor defect was apparently enough to doom a young person's chances, for as one physician of the period put it, the deformed 'are destined to drink from the dregs' of life's cup. Those struck or 'branded' (literally: *frappés d'un cachet*) by deformity, he said, were condemned to live as 'social pariahs'. 'I have a friend,' this doctor went on, 'who told me that myopia had destroyed his life: *"I'm nothing,"* he said, *"because I am good for nothing"*.'[7] We might recall, in this context, that one of Balzac's capricious heroines spurns a whole series of suitors on the basis of their physical defects – one is myopic, another limps, almost all of them lack the natural grace that distinguishes a man of quality.[8] Ambitious bachelors of the age were no less discriminating: the ruthless social climber Rastignac, for example, confesses to a friend that he has had to break off a romantic alliance because the object of his affections has a deformed foot. 'If people found out,' he says, 'I'd look ridiculous.'[9] With the rise of a romantic conception of marriage (imagined as a 'natural union' of mutually attracted partners), came heightened concern for the social significance of bodily flaws and defects. The correlation of moral and physical deformity was accepted as a self-evident truth by 'upright' bourgeois citizens, and the correction and prevention of physical defects in their offspring was perceived not just as a duty, but even as a privilege. Enrolling a child in a programme of physical rehabilitation (or an institute of physical 'education') conferred a certain benefit to the family as a whole; both the patient and his or her parents were seen to be participating in an exercise as morally elevating as it was socially uplifting. In short, there were significant new social and economic incentives, especially within the middle class, to seeking orthopaedic care.

These same years were marked by an economic crisis within the urban medical population: a surfeit of licensed practitioners produced a stagnation in honoraria.[10] In other words, the cost of care began to stabilize, and the profit-margin of private practice began to diminish. Recent graduates of the medical faculty had a considerable incentive to explore and exploit new markets, including those traditionally dominated by unlicensed practitioners. Orthopaedic medicine was one of these. As young physicians began to explore specializations outside of the institutional mainstream, the medical establishment began to broaden the scope of its regulatory activities, and even to investigate the possibility of integrating some formerly marginal practices within the more easily policed environment of the public hospital.

These, then, are the broad outlines of the context for orthopaedic medicine's remarkable growth in the early decades of the nineteenth century. I would like now to take a closer look at the ways in which the commercial success of orthopaedic practice contributed to the discipline's professionalization, and institutionalization in the public sector – that is, how the value of orthopaedic medicine was actually negotiated by those within, and without, the medical community. Here, we run into something of a paradox, for the 'high' and 'low' cultural productions of this period seem to suggest that orthopaedic medicine was remarkable only for its failings. Social satirists of the July Monarchy lampooned the public's guileless enthusiasm for 'one-act' cures and orthopaedic panaceas. In a comic engraving of the period a gnarled but charming gallant declares: 'Yes, my dear, yes, I, the Count of Hump-de-Back used to have a hunchback, a terrible hunchback, before orthopaedic medicine came to my rescue, and you see, now [that] the cure's complete, I'm straight as an "I" ' .

Well, perhaps not quite. But the credulous Count obviously wasn't the only one to be taken in by an unscrupulous operator. In a caricature published in 1841 a cross-eyed *naïf* proudly declares himself to be cured of a previous ocular defect. 'Now that I've been operated upon,' he says, 'I can see straight. It's really transformed me, no?' 'Oh absolutely,' his friend replies: 'as I recall, you were wall-eyed before.' There is a pun here that does not come across in translation: the term *louche* refers both to a visual impairment (strabismus), and to a lack of moral integrity, or probity. It was the impossibility of curing either defect that prompted a character in Flaubert's novel, *Madame Bovary*, to declare that operations for strabismus, like operations for club-foot or spinal deviation, were 'monstrous inventions' – fantasies that 'should be outlawed by the government'.[11]

Some physical defects, it seems, simply defied correction. Others had a kind of cachet or prestige that was, arguably, worth preserving. Consider, for a moment, the figure of the hunchback. The first to come to mind, of course, is Quasimodo – the 'half-mad' hero of Victor Hugo's *Notre Dame de Paris* (1831). But there are many others. At least two figure prominently in Balzac's *Comédie humaine* – the noble and self-sacrificing Butscha (in *Modeste Mignon*) and the equally heroic Joséphine Claës, in *La recherche de l'absolu*. Flaubert devoted five lines to the hunchback in his *Dictionary of Received Ideas*: 'touching their hump', we learn, 'brings good luck'. They are also 'exceptionally witty', and 'expert at riding whores'; perhaps for this reason, they are also 'always sought after by lascivious women'.[12] In short, they are not lacking in qualities – all of them associated, in one way or another, with their deformity.

One has to wonder, in this context, how the *tailleur des bossus* stayed in business at all. In fact, however, the 1820s and 1830s coincided with a veritable boom in the creation of private orthopaedic establishments (Fig. 6). Several were in operation in and around Paris by the time Georges Duval's comedy opened, in March of 1826. One of them, located in the outlying village of Chaillot, was run by Dr Vincent Duval (1795–1876), a recent graduate of the Paris Faculty of Medicine whose entrepreneurial aims coincided – fortuitously, it seems – with those of the fictional *Tailleur des Bossus*.[13] Like most of his colleagues and competitors, Dr Duval relied on the use of mechanical extension beds to treat scoliosis and other skeletal defects.[14] The vogue for such devices was so great that some enterprising orthopaedists envisaged

Fig. 6   Orthopaedic institute, c.1840. A crowd of disabled city-dwellers (distinguished by their bourgeois garb) flock to the gates of a suburban orthopaedic institute, seeking treatment. Scoliosis and club-foot are among the most visible deformities pictured here; not coincidentally, these were the conditions for which orthopaedic specialists claimed to offer the most dramatic and transformative cures. The illustration was designed to accompany a verse lampooning specialists and their commercial success within the contemporary medical marketplace. ('Les spécialités', in François Fabre, *Némésis médicale illustrée, recueil de satires, rev. et corr. avec soin par l'auteur; contenant 30 vignettes dessinées par M. Daumier, et gravées par les meilleurs artistes, avec un grand nombre de culs-de-lampe, etc.* [Paris: Bureau de la Némésis médicale, 1840, p. 67].)

implementing their use on a grand, even 'industrial' scale. In 1825, Duval's father-in-law (and business partner), Guillaume Jalade-Lafond, presented the Academy of Medicine with a design for a mechanism that could be used to operate several beds simultaneously.[15] The same year, Charles-Amédée Maisonabe, the director of an orthopaedic institute on the boulevard Montparnasse, presented the Academy with an extension bed of his own invention. The principles of its design and construction were diametrically opposed to those of Jalade-Lafond's machine, but it was built with the same end in view – namely, cornering the lucrative (and still expanding) market for orthopaedic traction machines.

Although disinclined to involve itself in what was clearly a commmercial dispute, the Academy of Medicine appointed a commission to evaluate the relative merits of each design, and the therapeutic benefits it procured. Six months later, the commission's *rapporteur*, Jean-Baptiste Thillaye, submitted an inconclusive report commending both inventors, but endorsing neither one of their prototypes.[16] Instead, Thillaye conveyed the commissioners' concern that even a well-designed and manufactured apparatus was susceptible to misuse, if it fell into the hands of an unqualified practitioner. In short, it was in the interest of medical professionals and patients alike to limit the use of devices like extension beds, which threatened the health and safety of the public, and the professional well-being of the physician, who risked both his honour and his honoraria in resorting to the use of these infernal machines.[17]

The extension bed was, indeed, a hellish invention – a kind of torture device, turned to therapeutic ends. Most of those in use in this period operated on the principle of 'permanent extension': the patient (usually an adolescent girl) was attached by her head and pelvis to a kind of Procrustean bed that was designed to stretch and forcibly straighten the spine (Fig. 7). Although widely used, this method was not regarded as being entirely satisfactory; the results it produced were usually temporary, and the tension that was applied to the spine was difficult to regulate. The patient, immobilized and often in great pain, was at the mercy of the bed's operator. There were, apparently, some rather awful accidents.[18] It was no doubt for this reason that the Paris hospital administration decided, in 1825, to look into the possibility of creating a special orthopaedic ward in one of the city's hospitals, where the use of these devices could be adequately supervised. Dr Péligot, one of the physicians who helped to evaluate extension beds for the Academy of Medicine, was put in

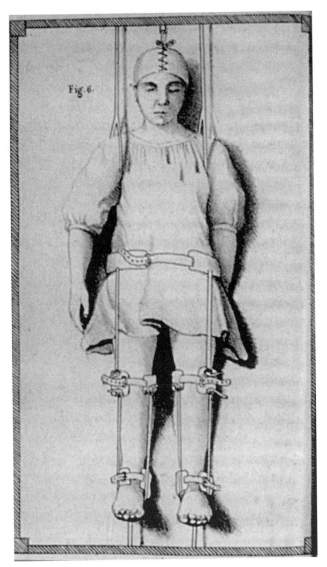

*Fig. 7*   Mechanical extension bed, c.1835. A mechanical extension bed of the type introduced by J. A. Venel in the 1770s. Similar devices (with certain modifications made by the German orthopaedist J. G. Heine) were still in use in Parisian orthopaedic institutes in the 1830s. (Fig 6. in François Louis Éduard Mellet, *Manuel pratique d'orthopédie, ou Traité élémentaire sur les moyens de prevenir et de guérir toutes les difformités du corps humain* [Paris: Rouvier et Lebouvier, 1835].)

charge of the affair.[19] More than a decade passed before any decisive action was taken.

In the intervening years, concern about the misuse of extension beds continued to spread. Claude Lachaise, a graduate of the Paris Faculty, published polemical works decrying what he described as the 'furore for mechanical redressment', and advocating a return to the gymnastic methods favoured by eighteenth-century physicians and philosophers.[20] Lachaise shared with Nicolas Andry (the physician who coined the term 'orthopaedia') and Jean-Jacques Rousseau a conviction that physical education was the best remedy for the general debility of mind and body produced by a sedentary social existence.[21] Like other hygienists of the period, he promoted exercise as a corrective therapy, one that would reinvigorate the weakened moral and muscular fibre of an indolent urban elite.[22] Contemporary critiques of 'permanent extension' cannily made use of this same rhetoric, rejecting in principle the therapeutic value of a technique that relied on the passive *decubitus* (or supine position) to lengthen or straighten the spine. Jalade-Lafond, for example, reprinted the Academy of Medicine's report of 1825 with an introduction and annotations defending his 'oscillatory' approach to extension as a kind of exercise, and rejecting devices (like Maisonabe's) that 'condemn the muscles to inactivity and [hence] to feebleness and atrophy'.[23]

In 1827, at the height of the backlash against extension devices, Dr Jalade-Lafond published a three-volume treatise on the importance of muscle tone in combating and correcting skeletal deformities.[24] The method of permanent extension, he said, was all wrong; what was needed was motion and muscular action that would stimulate the 'vital properties' and bring about a natural redressment. Jacques-Marie Delpech (1777–1832), the founder of a large orthopaedic institute in Montpellier, preached a similar gospel in his work on *Orthomorphia*, published in 1828.[25] Most deformities, he claimed, were due to muscular atrophy, or degeneration; the best remedy was a combination of gymnastic exercise and extension that restored form by restoring function. In place of the Procrustean extension bed, his institute featured an array of exercise equipment designed to appeal to a bourgeois clientele that associated mobility with self-sufficiency and social success.[26] Ladders, balances and self-propelled chariots were among the tools used to strengthen and 're-educate' the bodies of the institute's young patrons, most of whom were expected to remain in residence for a year or more.[27]

One of Delpech's disciples, the physician Charles-Gabriel Pravaz (1791–1853), brought similar devices within reach of Parisian social-climbers when, in 1826, he established an orthopaedic facility at a girls' boarding-school on the rue de Bellefonds, in Passy.[28] Eight years later, with the backing of a young colleague named Jules Guérin, Pravaz expanded his operations and relocated to the grounds of an elegant estate adjacent to the Bois de Boulogne.[29] The grandly named *Institut orthopédique de Paris* occupied two stately residences and an adjoining park known to all as the Château de la Muette, a *demeure galante* of the Bourbon monarchy that was expropriated by the Revolutionary government and then sold, in 1818, to a wealthy Parisian businessman (the famed piano manufacturer Sébastien Erard), who leased the property to Pravaz. The Institute admitted young men and women – separately housed on opposite ends of the park – for a stay of no less than three months, and offered, in addition to a daily regimen of exercise and individually regulated extension, a specially designed course in 'positive education' that included instruction in French, history, geography and arithmetic.[30] Supplementary lessons in drawing, painting, music and foreign languages were provided for an additional fee.[31]

Within a year of its opening, Pravaz and Guérin's institute, the largest establishment of its kind in Paris, was widely acknowledged to be the finest and most successful orthopaedic institution in all of France.[32] A delegation of doctors from the Academy of Medicine inspected the facility, and were unanimous in approving its design and operation.[33] Copies of the institute's illustrated prospectus circulated widely. One, accompanied by a letter from Pravaz and Guérin, reached the Academy of Sciences in February of 1835.[34] News of this altogether unremarkable event – the Academy received unsolicited prospectuses and testimonials from countless medical entrepreneurs and instrument-makers in this period – appeared, immediately, in the pages of the *Gazette médicale de Paris*, a journal owned and edited by Guérin himself.[35] For the rest of the year, the paper featured regular reports and editorial commentaries on the state of orthopaedic medicine in the capital; the *Gazette* served, in effect, as a bully pulpit, from which Guérin could expostulate on the evils of 'permanent extension', and promote his own views on the muscular origin of skeletal deformities.

It is not hard to see that Guérin's privileged position as the publisher of a widely read medical journal enabled him to engage in a kind of indirect advertising that directly benefited his own commercial and professional interests.[36] Still, it would be misleading to suggest that Guérin's control of the new periodical media is enough to explain why the medical establishment, which had for years opposed the proliferation of private orthopaedic institutes, so unambiguously endorsed the principles and practices of the new *Institut orthopédique*. Part of the explanation for this apparent about-face has to do with debate surrounding the use and abuse of mechanical traction devices. Ten years after the imperfectly resolved dispute between Guillaume Jalade-Lafond and Charles Maisonabe, this issue was once again brought before the Academy of Medicine, when an unlicensed orthopaedist (a certain Mr Hossard) from the city of Angers arrived in the capital, and began distributing handbills on the Pont-Neuf, claiming that the Academy had officially sanctioned the use of his new 'inclination lever belt' in the treatment of spinal deviations.[37] (See Fig. 8.)

The Academy was scandalized by this affair, which reawakened old fears about the abuse of official endorsements, and the misuse of orthopaedic machines. Guérin, who reported regularly on the Academy's meetings in the *Gazette médicale*, seized the opportunity to condemn the unregulated practice of orthopaedic medicine, and did not hesitate to identify the practitioner from Angers as a charlatan, and *Crispin*.[38] For months, a battle over the best treatment for spinal deviations raged at the Academy, and was recorded on the pages of the *Gazette médicale*. Finally, in September of 1835, Guérin addressed a letter to the Academy of Medicine, in which he accused the orthopaedist from Angers of having 'faked' cures that the Academy had previously accepted as genuine.[39] The accused responded by filing suit against Guérin.[40] New anti-libel legislation was in his favour, and he won the case. But the real victor was Guérin, who succeeded in galvanizing support within the medical profession for a stricter application of controls against the unlicensed practice of orthopaedic medicine, and, above all, the abuse of mechanical devices. The gains made in 1835 were ultimately ratified by legislation that restricted the legal use (or 'prescription') of mechanical extension devices to licensed physicians, thus consolidating the medical establishment's control over a range of competing therapeutic practices (including bone-setting, truss-making, drug preparation and midwifery) and personnel.[41]

In the years that followed, Guérin continued to proselytize in favour of ortho-gymnastic methods of redressment, and the theory of muscular atrophy upon which they were based. He inscribed his critique of extension within a larger argument about the organization of the contemporary medical profession, and the principles that directed its practice. Guérin was an advocate of medical 'eclecticism' who argued that blind faith in monolithic doctrines like the theory of irritation

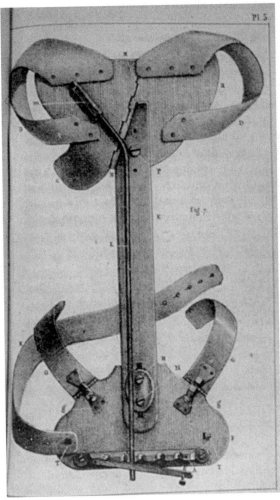

*Fig. 8* Inclination lever belt, c.1835. Inclination lever belt designed by F.-L.-E. Mellet and used in his orthopaedic practice during the 1830s. Introduced as an improvement upon traditional orthopaedic corsets, which compressed the whole thoracic region, Mellet's belt was designed to reduce pressure on the chest while forcibly extending the spine. The action of the device mimicked a chiropractic manipulation in which pressure was simultaneously applied to opposite sides of the spine to straighten the back; Mellet recommended that the belt be used in combination with gymnastic therapy to strengthen the muscles supporting the spine. (Fig. 7.1 in F.-L.-E. Mellet, *Manuel pratique d'orthopédie, ou Traité élémentaire sur les moyens de prévenir et de guérir toutes les difformités du corps humain* [Paris: Rouvier et Lebouvier, 1835].)

(which was used by some, on the authority of Glisson, to justify forcible reduction and mechanical extension) put the practitioner, the patient and the profession as a whole at risk. This was the thesis of a memoir he presented to the Paris Academy of Sciences in 1833, a year after a devastating cholera epidemic had revealed the medical profession's powerlessness to protect even the most highly placed members of the social and political elite.[42] Guérin was awarded the prestigious Prix Monthyon for this work, which brought him to the attention of the intellectual community at large.

In 1837, the Academy of Sciences awarded Guérin another major prize, for a sixteen-volume work on the origin and treatment of human skeletal deformities.[43] Here, Guérin laid out in detail his views on the etiology of spinal deviation, and the other defects he treated at his establishment in Passy. Again, he insisted on the role of muscular degeneration in their production, and the importance of exercise in their treatment. The Academy's prize commission published a report praising the justice of Guérin's far-reaching views on the organization of the animal economy, noting, in particular, the interest of his allegation that physical activity was essential to preserving (or restoring) the structural symmetry and stability of the organism. Guérin reprinted the report, and had it bound with a new prospectus for the *Institut Orthopédique*, which had recently come under his exclusive control.[44] Pravaz, meanwhile, relocated to Lyon, where he assumed responsibility for a flourishing 'branch location' of the Paris Institute.[45]

Guérin's increasing celebrity in the scientific and medical world was important to ensuring the commercial success of his therapeutic methods; it was also important to protecting his own professional reputation, as a practitioner who worked on the insitutional margins of the contemporary medical establishment. Guérin was, in effect, operating in two different markets: one defined by the interests of the medical establishment, increasingly concerned in this period by the proliferation of marginal practices and private institutions; and another defined by the demands of an expanding bourgeois clientele that sought out and patronized those very same establishments. It was by negotiating a place for himself between these interests, that he succeeded in legitimating a discipline that had for centuries been practised outside the insitutional mainstream.

Guérin's greatest triumph came in 1839, when the Paris hospital administration approved the creation of a special twelve-bed orthopaedic ward at the Hôpital des Enfants Malades, where Guérin was authorized to lecture on muscular retraction, and provide clinical instruction in tenotomy, a technique he had begun to argue was the only rigorously scientific method of treating skeletal deviations.[46] The method was not new, and Guérin was not the only one to use it, but he was the only one to recognise that it provided the key to a general etiological theory of deformity that would unify the practice of orthopaedic medicine.[47] It involved sectioning (i.e. cutting) the tendons that connected pathologically altered muscles to the skeleton; restoring the body's alignment by severing the connection between bone and muscle.[48] Although continuous with the theory of myogenic atrophy that Guérin had used to justify the gymnastic methods used at his Institute, this new therapeutic gesture made it possible, for the first time, to justify the public institutionalization of orthopaedic medicine – for, reconfigured as a surgical speciality, the discipline entered the realm of 'hospital medicine'.

More than any other orthopaedist of his generation, Guérin knew how to capitalize on the 'moment' in which he operated. His pragmatic adaptation to the new professional and social realities of the July Monarchy enabled him to mediate effectively between the professional and the commercial marketplace, and to generate tangible (and intangible) profits in each of them. He was, unquestionably, the motor force behind the disciplinary development of orthopaedic medicine in France. Ironically, Guérin's own career within the discipline he had helped to legitimate and institutionalize was short-lived. Within five years of his appointment at the Paris Children's Hospital, he became the target of vituperative attacks by rival orthopaedists (Malgaigne and Bouvier, in particular), who published diatribes against his methods, and impugned his professional scruples in the popular periodical press, and at meetings of the Academy of Medicine.[49] In 1843 Malgaigne, once a collaborator on the *Gazette médicale*, rebuked Guérin for self-publishing (and widely distributing) a report that purportedly 'proved' the success of his methods.[50] Guérin responded just as his old nemesis Hossard had, ten years earlier: he filed suit.[51] This time, however, the Academy of Medicine rallied to the side of his opponents, and he lost the court battle.[52] Shortly thereafter, in spite of a glowing statement (signed, this time, by a panel of highly respected physicians) on the results achieved in the orthopaedic ward, Guérin was relieved of his duties at the hospital, and the Institut Orthopédique de Paris was closed.[53] The year was 1849 – the social and political context of the July Monarchy was over: his 'moment' had passed.

There is a famous episode in *Madame Bovary* (1857) in which the hapless *officier de santé*, Charles Bovary, undertakes an ambitious project to cure a local stable-boy of his debilitating club-foot. Bovary is goaded into performing this supposedly philanthropic act by his wife, who longs to see Charles as a heroic man of action, and by the village pharmacist, who hopes that the publicity surrounding the event will benefit his own business. While Bovary, who has never before performed the procedure, prepares by studying a surgical manual, the pharmacist Homais busies himself with persuading the stable-boy (Hippolyte) that it is in his best interest to undergo the operation.[54] The procedure, Homais says, is practically painless, and is certain to 'make a man' of the boy – once relieved of his 'hideous limp', Hippolyte is certain to become more agile, virile and attractive to the opposite sex.[55] Hippolyte, who is over twenty-five, and still unmarried, is entranced by Homais's words, and when he learns that the operation will 'cost him nothing', he submits readily to the action of Bovary's brand new surgical knife.

In the end, as we all know, the bungled operation ends up costing both the stable-boy and his benefactor a good deal. Five days after Bovary has, in accordance with the instructions outlined in his manual, sectioned the Achilles tendon and immobilized Hippolyte's foot in an eight-pound brace, gangrene sets in. A licensed physician (Dr Canivet) is summoned to Yonville from Neufchâtel; he declares the case to be hopeless, and proceeds to amputate Hippolyte's leg from the hip. Bovary, humiliated and wracked with guilt, is forced not only to pay the price of a highly publicized failure, but also that of a costly prosthetic device, which, ironically, turns out to be of almost no use to his patient. The stable-boy finds the prosthesis too ostentatious, and wears it just once, at the funeral of Emma Bovary; the clatter it makes as he crosses the floor of the church sounds to the grief-stricken Charles like a reproach, a reminder of his failings in life and in love.

Flaubert does not tell us why Bovary's attempt to cure Hippolyte of his disability went wrong, and he does not need to. The operation was doomed from the outset – not because of any intrinsic difficulty or danger in the procedure, which was used with great success by a number of contemporary practitioners (including Guérin, and Vincent Duval – the original *Tailleur des bossus*), but because Charles Bovary was destined to fail in any ambitious enterprise. Flaubert used the episode to dramatize the futility of Bovary's efforts to transcend or transform his own (or anyone else's) destiny. As the cynical Dr Canivet scoffs, 'Correct a clubfoot! One might as well try to straighten a hunch-back!'[56] This is, in fact, exactly what the exasperating duo in Flaubert's *Bouvard et Pécuchet* (1881) attempt to do; like Charles Bovary, they rely on a manual to guide their practice, and, like him, they fail. The cure-all camphor cataplasms they apply are no more effective than Bovary's *ténotome* in 'righting' or fighting the order of things.

Like Georges Duval before him, Gustave Flaubert had his doubts about the quick-fix remedies of orthopaedists and other 'operators' on the contemporary social stage. But in the years that separated the *Tailleur des bossus* and Charles Bovary, the comic register had shifted from light-hearted farce to bitter satire, as the social and economic transformations embodied by the rise (and fall) of entrepreneurs like Jules Guérin made it clear that the liberating rhetoric of self-sufficiency and social promotion (*enrichissez-vous!*) had failed to restore harmony, symmetry or stability to post-Revolutionary France. It was surely the impossibility of obtaining such an outcome by means of any miracle cure that made Flaubert conclude that the hunchback's greatest defect was, in fact, his greatest virtue.

## Notes

1. *Le tailleur des Bossus; ou, L'orthopédie, contrefaçon en un acte et en vaudeville, par MM. Georges Duval, Rochefort et \*\*\* représentée pour la première fois à Paris, sur le théâtre des Vaudeville, le 20 mars 1826* (Paris: Barba, 1826).

2. On the professionalization (and institutionalization) of medical specialities in France during this period, see George Weisz, 'The Development of Medical Specialization in Nineteenth-Century Paris', in Ann La Berge and Mordechai Feingold (eds), *French Medical Culture in the Nineteenth Century* (Amsterdam/Atlanta: Rodopi, 1994), pp. 149–88 and especially pp. 154–6; Erwin H. Ackercknecht, *Medicine at the Paris Hospital, 1794–1848* (Baltimore: Johns Hopkins University Press, 1967), pp. 163–80; and, with special reference to the development of social medicine and rehabilitative care, Dora B. Weiner, *The Citizen-Patient in Revolutionary and Imperial Paris* (Baltimore: Johns Hopkins University Press, 1993), pp. 184–90, 225–46, and *passim*).

3. Existing historical scholarship on the development of orthopaedic medicine in France has, by and large, focused on the evolution of therapeutic techniques (especially the rise of surgical orthopaedics), without regard to the changing social and economic context of medical practice in this period. Cf. Leonard Peltier, *Orthopedics: a History and Iconography* (San Francisco: Norman, 1993), especially pp. 195–209 (on scoliosis); and David Le Vay, *The History of Orthopaedics: an Account of the Study and Practice of Orthopaedics from the Earliest Times to the Modern Era* (Park Ridge, NJ: Parthenon, 1990), esp. pp. 238–59.

4. Anne Martin-Fugier, *La vie élégante, ou la formation du Tout-Paris, 1815–1848* (Paris: Fayard, 1990).
5. On the normalization of aesthetic ideals in middle-class dress of this period, see Philippe Perrot, *Fashioning the Bourgeoisie: a History of Clothing in the Nineteenth Century* (Princeton: Princeton University Press, 1994).
6. A leading hygienist, the physician J.-H. Réveille-Parise (1782–1852) reported on the case in a *feuilleton* published in the *Gazette médicale de Paris*, vol. 3 (27 June 1835), 401–5, esp. p. 405. The anecdote was related in the context of a testimonial vaunting the advantages of the 'dynamic' therapeutic regimen employed at a private orthopaedic clinic (the Institut Orthopédique de Paris) run by Réveille-Parise's friend, Dr Jules Guérin – also editor, owner and publisher of the *Gazette médicale*.
7. *Gazette médicale* (1835), 404–5. It should be remembered in this context that the practice of branding convicted criminals had only recently been abolished in France (c. 1832). By identifying the correction of deformities with the social rehabilitation of prisoners and 'misfits', Réveille-Parise consciously inscribed the practice of orthopaedic medicine within the liberal political and economic agenda of the July Monarchy's middle-class elite: once freed of physical restraints, patients and prisoners were expected to regain both dignity and social mobility.
8. Though 'rich and beautiful enough to be able to choose [a husband] from amongst the world's princes', the heroine of *Le Bal de Sceaux* (Emilie de Fontaine) rejects, on the basis of their physical defects, all of the bachelors presented to her.
9. See chapter 3 ('La femme sans coeur') in Honoré de Balzac, *La Peau de chagrin*.
10. George Sussman, 'The Glut of Doctors in Mid-Nineteenth-Century France', *Comparative Studies in Society and History*, 19 (1977), 293–303.
11. It is Dr Canivet, the physician summoned to Yonville following Charles Bovary's disastrous experiment with tenotomy, who offers this judgement. Flaubert, it should be remembered, was the son of a provincial physician (Achille-Cléophas Flaubert, surgeon at the Hôtel-Dieu in Rouen and a member of the Paris Academy of Medicine), and well-informed about contemporary medical practice. J. Normand and T. Irles, 'Gustave Flaubert et la médecine', in *Conférences d'histoire de la médecine* (Lyon: Fondation Marcel Mérieux, 1992), pp. 21–32 (and esp. pp. 27–30); René Dumesnil, *Flaubert et la médecine* (Paris, 1905).
12. Gustave Flaubert, *Dictionnaire des idées reçues*, in *Oeuvres complètes*, II (Paris: Seuil, 1964), p. 304.
13. Originally located near the porte Maillot (on the northern edge of the Bois de Boulogne) Duval's clinic moved several times within the village of Chaillot before relocating, during the Second Empire, to Neuilly. The clinic continued to operate after Vincent Duval's death (in 1876), when the business was inherited by his son, Emile. On the physical arrangement of Duval's facilities in Chaillot, see [Anon.], 'Souvenirs du vieux Chaillot: La maison du docteur Duval', *Bulletin de la Société d'Histoire d'Auteuil et du Passy*, 6 (1916), 15–16.
14. Adolphe Milli, the director of a private orthopaedic establishment in Chaillot, is generally credited with having introduced the extension bed to

Paris practitioners, around 1822. Milli had been successfully treated for spinal curvature with a device like this at J. G. Heine's institute in Würzburg, Germany, and began importing the beds to France upon his return. On Milli's role in the promotion of extension beds, see Joseph-François Malgaigne, *Leçons d'Orthopédie professées à la Faculté de Médecine de Paris* (Paris: Delahaye, 1862), p. 393; Bruno Valentin, Geschichte der Orthopädie (Stuttgart: Thieme, 1961), pp. 42, 199. On Heine's orthopaedic institute, see [Anon.] *Notice sur l'institut orthopédique du Dr. Heine, à Wurtzbourg* (Paris, n.d.). Milli's clinic was purchased (c. 1824) by Dr Henri-Sauveur Bouvier, who at first championed the therapeutic benefits of extension, but later acknowledged a role for 'dynamic' or gymnastic techniques in the treatment of skeletal deformities.

15. Guillaume Jalade-Lafond, *Exposé succinct des moyens mécaniques oscillatoires imaginés et employés pour rémédier aux déviations de la colonne vertébrale et autres vices de conformation, suivi d'un rapport fait à l'Académie Royale de Médecine, par MM. Breschet, Husson, etc., pour lui faire connaître les avantages ... des mécaniques oscillatoires orthopédiques de M. Jalade-Lafond* (Paris: Boiste, 1825).

16. Thillaye himself was a recognized expert in the design and use of trusses, bandages and other medical supports, especially in the treatment of injuries sustained on the battlefield, viz., Jean-Baptiste-Jacques Thillaye, *Traité des bandages et appareils, à l'usage des chirurgiens des armées* (Paris: Crochard, 1809).

17. See Thillaye's remarks in the official *rapport* appended to Jalade-Lafond, *Exposé*, p. 33.

18. These were due, in part, to pathological changes in the vertebrae which rendered them less flexible, and more apt to break when the patient was placed in a traction device. The development of longitudinal traction machines was in fact motivated by a desire to *reduce* stress on the spine; earlier orthopaedic apparatus had been designed to 'flatten' gibbous protrusions by direct pressure (of a metal plate or other unyielding plane) regulated by a common screw.

19. Archives de l'Assistance Publique, Conseil Général des Hospices, arrêté 16 mai 1825 (no. 41717), ordering Dr Péligot to draft a report on contemporary orthopaedic practices, in preparation for the creation of a specialised ward in one of the Paris hospitals. See also Thillaye's comments in Jalade-Lafond, *Exposé*, pp. 12, 16.

20. Claude Lachaise, *Précis physiologique sur les courbures de la colonne vertébrale; ou, Exposé des moyens de prévenir et de corriger les difformités de la taille, particulièrement chez les jeunes filles, sans le secours des lits mecaniques à extension* (Paris: Villeret, 1827), p. xi. Under the pseudonym 'C. Sachaile de la Barre', Lachaise published a 'scientific and moral' evaluation of his peers in the Paris medical community, *Les médecins de Paris jugés par leurs oeuvres; ou, Statistique scientifique et morale des médecine de Paris* (Paris, 1845). Of the nearly 1500 practitioners profiled in that work, Lachaise identified only seventeen orthopaedists, including some (like Ferdinand Martin, a truss- and instrument-maker at the famous Invalides military hospice) who were not physicians, but mere *officiers de santé*.

21. Andry introduced the term in *L'orthopédie ou l'art de prévenir et de corriger dans les enfans les difformités du corps* (Paris: la veuve Alix, Lambert &

Durand, 1741) translated as *Orthopaedia; or, The art of correcting and preventing deformities in children*, 2 vols (London, 1743).
22. Claude Lachaise, *Topographie médicale de Paris* (Paris: Baillière, 1822), esp. pp. 278–301 on the therapeutic benefits of physical exercise.
23. G. Jalade-Lafond, *Exposé*, pp. 5–10.
24. Guillaume Jalade-Lafond, *Recherches pratiques sur les principales difformités du corps humain, et sur les moyens d'y remédier* (Paris: Baillière, 1827–9).
25. Jacques-Mathieu Delpech, *De l'Orthomorphie, par rapport à l'espèce humaine: ou Recherches anatomico-pathologiques sur les causes, les moyens de prévenir, ceux de guérir les principales difformités et sur les véritables fondemens de l'art appelé orthpédique* (2 vols, atlas) (Paris: Gabon, 1828–9).
26. Delpech defended the originality of his 'gymnastic' methods, which he claimed to have used in treating patients at his clinic as early as 1825. Delpech, *De l'Orthomorphie* (1828), p. 181. See also Leonard F. Peltier 'The "Back School" of Delpech in Montpellier', *Clinical Orthopaedics and Related Research*, 179 (October, 1983), 4–9.
27. The facilities, grounds and devices in use at Delpech's institute are described and illustrated in the atlas to *De l'Orthomorphie* (1828–9).
28. Though best remembered by historians as the inventor of the hypodermic syringe, Charles-Gabriel Pravaz first distinguished himself in the field of orthopaedic medicine by introducing to France the method of 'local extension' promoted by London physician John Shaw. Charles-Gabriel Pravaz, *Méthode nouvelle pour le traitement des déviations de la colonne vertébrale, précédée d'un examen critique des divers moyens employés par les orthopédistes modernes* (Paris: Gabon, 1827). A report on Pravaz's method and apparatus was read before the Paris Académie de Médecine in 1828.
29. [Anon.], *Institut Orthopédique de Paris, pour le traitement des difformités de la taille et des membres, chez les personnes des deux sexes; dirigé par MM. les docteurs Pravaz et Jules Guérin, au Château de la Muette, à Passy, près le bois de Boulogne* (Paris, 1834).
30. For young men, special instruction in Latin, mathematics and 'elements of the Natural Sciences' was provided at no additional cost. [Anon.], *Institut Orthopédique de Paris* (1834), p. 7.
31. In addition to courses in French and Italian song and basic piano technique, residents of the Institut could pursue advanced keyboard training with a certain M. Chopin. It should be noted, in this regard, that the expected social accomplishments of upper- and middle-class women (drawing, painting and keyboard performance) were singled out for critique by orthopaedists like Shaw and Delpech. The famous 'counter-weight column' devised by Delpech to maintain correct posture at the keyboard or drawing table was, in fact, copied (as Delpech himself acknowledged) from Shaw's design.
32. See e.g. H.W. Berend, 'Die orthopädischen Institute zu Paris', *Magazin für die gesamte Heilkunde*, 59 (1842), 496; Valentine Mott, *Travels in Europe and the East* (New York: Harper, 1842), p. 55.
33. *Gazette médicale* (1835), 405.
34. Reported and summarised in the *Gazette médicale*, 3 (7 February 1835), 89.
35. The recently created Academy of Medicine (established in 1820) played an increasing important role in this period in endorsing different therapeutic regimes and devices. Cf. Matthew Ramsey, 'Academic Medicine and Medical

Industrialism: the Regulation of Secret Remedies in Nineteenth-Century France', in La Berge and Feingold, *French Medical Culture*, pp. 25–78.

36. As Malgaigne (who founded two medical journals himself) observed, Guérin relied on the *Gazette médicale* to publicize his medical and scientific views, especially before he gained a seat (and a voice) at the Academy of Medicine, in 1842. (Malgaigne was not elected a member until 1846.) See Malgaigne's comments in the *Journal de Chirurgie*, 1 (1843), 20.

37. On the controversy surrounding Hossard's *ceinture à lévier*, see J.-F. Malgaigne, *Leçons d'Orthopédie*, pp. 406–14. See also M. J. Hossard, *Traitement des déviations de la taille sans lits mécaniques, système d'inclinaison employé à l'établissement orthopédique d'Angers, Maine et Loire* (Angers, 1853).

38. See Guérin's report on Hossard's activities at the Academy of Medicine in February of 1835, published in the *Gazette médicale de Paris*, 3 (21 February 1835), 123–4, continued in his editorial *feuilleton* ('Lettre médicale sur Paris'), *Gazette médicale de Paris*, 3 (28 February 1835), 129–31. Further reports (and an exculpatory *réclamation* by Hossard) appeared in the *Gazette médicale*, 3 (1835) in the months that followed: see p. 141 (February); pp. 161–3 (March); and pp. 593–6, 605–7 and 622–3 (September). See also [J.-F. Malgaigne], 'Les spécialités en police correctionelle – Procès de M. J. Guérin', *Revue médico-chirurgicale*, 3 (January 1848), 64. The epithet 'Crispin' is an allusion to a character (Crispinus, an ambitious but ignorant critic of the poet Horace) in Ben Jonson's *Poetaster* (1601); in French comedy, the term is synonymous with arrogance and self-delusion, typically personified by an insolent valet, as in A. R. Le Sage's *Crispin, rival de son maistre* (Paris, 1737) or N. Le Breton de Hauteroche's *Crispin médecin* (Paris, 1674) – both revived in new stage productions around 1808.

39. Jules Guérin, *Mémoire sur les déviations simulées de la colonne vertébrale et les moyens de les distinguer des déviations pathologiques, présenté a l'Académie Royale de Médecine le 31 Mai 1836* (Paris: Gazette Médicale, 1838). See also the reports published in the *Gazette médicale de Paris*, 3 (1835): 606–7, 622–3. The standard of proof applied in determining the success (or failure) of a specific therapeutic technique varied, where orthopaedic treatment was concerned. In Hossard's case, plaster casts of the original deformities (spinal deviations) were compared to the spinal columns of the same patients, following a course of treatment with the *ceinture à levier* (*l'appareil Hossard*). Delpech used the same technique (in *De l'Orthomorphie*) to document his success in treating scoliosis, but the results of his treatment were evidently not presented to any official tribunal (like the Academy of Medicine) for assessment.

40. Malgaigne characterized the dispute between Hossard and Guérin as a 'quarrel between patent peddlars' in a scabrous review article, 'Les spécialités en police correctionelle – Procès de M. J. Guérin', *Revue médico-chirurgicale*, 3 (January 1848), 64.

41. The new legislation regulating medical education and practice enacted under the Second Republic was reviewed in Malgaigne's *Revue médico-chirurgicale*, 3 (January 1848), 55–63. See especially Title IV, Article 20 (regulating the distribution of orthopaedic devices), and Title VIII, Article 6 (naming the penal consequences of violating Title IV, Art. 20), pp. 59, 62.

Malgaigne noted that this Article was revised several times, to establish a distinction between the 'profession to exercise' and the 'act of exercising'. See also Title V, Article 23 (prohibiting the use of handbills or posters to advertise medical treatments or remedies), p. 60.

42. Jules Guérin, *Mémoire sur l'éclectisme en médecine, précédé d'un rapport fait à l'Académie royale de médecine de Paris* (Paris: *Gazette médicale*, 1831); Guérin, *Examen de la doctrine physiologique, appliquée à l'étude et au traitement du choléra-morbus, suivie de l'histoire de la maladie de M. Casimir Périer* (Paris, 1832). See also Ackercknecht, *Medicine at the Paris Hospital*, pp. 101–13.

43. The prize question (*Déterminer par une série de faits et d'observations authentiques, quels sont les avantages et les inconvéniens des moyens mécaniques et gymnastiques appliques aux difformite du systeme osseux*) had been set by the Academy seven years earlier. Competitions held in 1830, 1832 and 1836 failed to satisfy the Academy's prize commission, which finally awarded Guérin the grand prize in 1837. Dr S.-H.-V. Bouvier (whose submission – an eight-volume memoir on the benefits of gymnastic exercise – was exactly half as long as Guérin's) was awarded a prize of 6000 francs. The history of the prize competition is summarized in the article 'orthopédie' published in Dechambre's *Dictionnaire encyclopédique des sciences médicales*, XVII, pp. 722–52.

44. [Anon.], *Institut Orthopédique de la Muette, pour le traitement des difformités de la taille et des membres, dirigé par M. le docteur Jules Guérin* (Paris, 1837), issued with [Académie Royale des Sciences], *Concours pour le grand prix de chirurgie relatif aux difformités du système osseux. Grand prix de 10,000 francs à M. Jules Guérin. Commission MM. Dulong, Savart, etc.; Double, rapporteur* (Paris, 1837).

45. Valentin, *Geschichte der Orthopädie*, p. 201. Pravaz's name is not mentioned in the 1837 prospectus for the Institut Orthopédique de Paris, but appears in a number of publications in Lyon, e.g.: Charles-Gabriel Pravaz, *Mémoire sur l'application de la gymnastique au traitement des affections lymphatiques et nerveuses, et au redressement des difformités: présenté à la Société de médecine de Lyon* (Lyon: Ch. Savy jeune, 1837).

46. Guérin's appointment to serve as head of the new orthopaedic service – made in the absence of any competitive examination, or *concours* – was widely criticized by contemporaries. Among the most vocal objectors was Dr Sauveur-Henri-Victor Bouvier (1799–1877), formerly responsible for the orthopaedic outpatient services of the Paris hospital system's central admitting office (*bureau central*), and the director of a private clinic in Chaillot. As early as 1835, Bouvier and Guérin were engaged in a priority dispute concerning the design and use of an extension device (what Guérin called his 'sigmoid extension apparatus') similar in principle to Hossard's 'inclination lever belt'. See e.g., *Gazette médicale de Paris*, 3 (21 November 1835), 747, (28 November 1835), 761–2 (12 December 1835), 796–8.

47. Both Duval and Delpech had used tenotomy successfully in the treatment of club-foot before Guérin began to generalize the procedure's use. Guérin first applied the technique (sectioning muscles, rather than tendons) to the correction of spinal deviations in 1839, and to the correction of strabismus – which he referred to as 'club-foot of the eye' – in 1841. J. Guérin, *Mémoire sur l'étiologie générale du strabisme lu à l'Académie Royale des Sciences, le 25*

*janvier 1841* (Paris: *Gazette médicale*, 1843) and Guérin, *Première mémoire sur le traitement des déviations de l'epine par la section des muscles du dos*, 2nd ed. (Paris: *Gazette médicale*, 1843). See also J. Guérin, *Mémoire sur l'étiologie générale des pieds-bots congénitaux, lu à l'Académie de Médecine, le 11 Decembre 1838* (Paris, *Gazette médicale*, 1838) and Guérin, *Mémoire sur l'étiologie générale des déviations latérales de l'épine, par rétraction musculaire active, lu à l'Académie royale des Sciences, le 23 septembre 1839* (Paris: *Gazette médicale*, 1840).

48. Bouvier was among those who critiqued Guérin's over-use of this procedure in the 1840s. See e.g. his comments at a meeting of the Paris Academy of Medicine (16 October 1848), reprinted in the *Revue médico-chirurgicale de Paris*, 4 (November 1848), 319–20. See also Bouvier's remarks on Guérin's use of myotomy in the treatment of spinal deviations, published in the *Gazette médicale*, 11 (15 July 1843), 454–6, and J.-F. Malgaigne, *Sur l'abus et le danger des sections tendineuses et musculaires dans le traitement de certaines difformités, mémoire adressé à l'Académie Royale des Sciences, le 5 fevrier 1844* and Malgaigne, *Mémoire sur la valeur réelle de l'orthopedie et spécialement de la myotomie rachidienne dans le traitement des déviations latérales de l'epine* (Paris: Baillière, 1845).

49. [J.-F. Malgaigne], 'Discussion sur la ténotomie', *Journal de Chirurgie*, 1 (January 1843), 19–25, 55–6.

50. J. Guérin, 'Relevé général du service orthopédique de l'Hôpital des Enfants', *Gazette médicale*, 11 (11 July 1843), 424. Of a total of 1394 cases admitted to the hospital, Guérin claimed to have 'completely cured' some 377, and 'improved' the situation of an additional 296. Twenty-five patients were reported to have died during the course of treatment (though not from any direct result of their treatment). An additional 619 patients were effectively excluded from Guérin's assessment on the grounds that their treatment had not been completed, or was never undertaken. For Malgaigne's critique, see 'Relevé général du service orthopédique de l'Hôpital des Enfants', *Journal de Chirurgie*, 1 (July 1843), 246, and 'De quelques illusions orthopédiques, à l'occasion du Relevé général du service orthopédique de M. J. Guérin', *Journal de Chirurgie*, 1 (August 1843), 257–65. Guérin's reported results were also challenged by Dr Charles Maisonabe, who addressed several critical letters on the subject to the editor of the *Gazette des hôpitaux* (François Fabre, also publisher of the satirical *Némésis médicale*). See correspondence published in the *Gazette des hôpitaux*, 5 (6 July 1843), 315–16, and 5 (15 July 1843), 318.

51. Guérin targeted the editors of two popular medical journals in which 'libelous' critiques of his work had been published (J.-F. Malgaigne of the *Revue médico-chirurgicale de Paris*, and Auguste Vidal de Cassis of the *Annales de la Chirurgie Française et Etrangère*), and a journalist, M. Henroz. Guérin sought (but did not win) a financial reparation of 20,000 francs, which was to be awarded to the Hôpital des Enfants Malades to support the services of the orthopaedic ward. See: 'Nouvelle simplification des discussions orthopédiques. – Procès intenté au *Journal de Chirurgie*', *Journal de Chirurgie*, 1 (1843), 321–2. [Jules Guérin], *Mémoire à consulter pour M. Jules Guérin, contre MM. Malgaigne, Vidal (de Cassis) et Henroz* (Paris: Malteste, 1844). See also Leonard Peltier, 'Guérin Versus Malgaigne: a Precedent for the Free

Criticism of Scientific Papers', *Journal of Orthopaedic Research*, 1 (1983), 115–18.

52. The trial came before the court in November 1843; the judge's ruling (against Guérin) was appealed the next year, but failed to overturn the original verdict, according to which all charges against Malgaigne were dismissed, and only modest penalties were assessed against Henroz and Vidal. In spite of this, Guérin continued to press for a retrial until 1846. Peltier, 'Guérin vs. Malgaigne', 117.

53. Malgaigne went so far as to imply that the text of the *rapport*, which offered fulsome praise for Guérin's methods and technique, had been written by Guérin himself, and then signed (but not read or critiqued) by the Academy's commissioners – all of whom, Malgaigne cynically observed, were friends and associates of Dr Guérin. [J.-F.] Malgaigne, 'Lettre à un chirurgien de province sur le Rapport addressé à M. le Délégué du gouvernement provisoire sur les traitements orthopédiques de M. le docteur Jules Guérin, etc.; par une commission composée de MM. Blandin, P. Dubois, Jobert, Louis, Rayer et Serres; président, M. Orfila', *Revue médico-chirurgicale*, 4 (October 1848), 249–58. See also Malgaigne's sarcastic send-up of an article by the astronomer and cleric l'Abbé Moigno, defending Guérin's claims to scientific integrity and originality. [J.-F. Malgaigne], 'A quoi servent certains grands prix de l'Institut', *Revue médico-chirurgicale*, 4 (November 1848), 322.

54. The surgical manual used by Bovary is identified by Flaubert as [Vincent] Duval's *Traité pratique du pied-bot* (Paris: Baillière, 1839). Vincent Duval had studied with Flaubert's father at the Paris Faculty of Medicine in the 1820s.

55. As Perrot notes, the highly polished shoes and finely fit slippers of July Monarchy fashions were designed to draw attention to the feet, to shape and model them into objects of desire. The eroticism of the well-formed foot is played to advantage in another of Flaubert's works, *L'Education sentimentale* (1869), in which the hero, Frédéric Moreau, experiences the first transports of 'possessing' his mistress (Madame Arnoux) when he spies her foot descending a stair. On the basis of his experience in treating some 105 cases of this deformity, F.-L.-E. Mellet concluded that young men were more susceptible to club-foot than were young women; see his *Manuel pratique d'orthopédie, ou Traité élémentaire sur les moyens de prévenir et de guérir toutes les difformités du corps humain* (Paris: Librairie des sciences médicales, 1835), p. 396.

56. As Malgaigne himself acknowledged, the traditional focus of orthopaedic medicine was the correction or straightening of crooked spines and feet, see e.g., J.-F. Malgaigne, *op. cit.* (1862), vi. Even after the speciality broadened its focus to include a range of joint and bone pathologies, club-foot and spinal deviations (the 'classic' locomotor disabilities) remained the focus of debate and discussion. In the 1860s, Malgaigne devoted the greatest part of his lectures at the Paris Faculty of Medicine to the etiology and treatment of scoliosis and club-foot. J.-F. Malgaigne, *Leçons d'Orthopédie*, pp. 105–74, 312–426. See also M. Pizetti, 'Histoire du traitement du pied-bot et de la scoliose', *The Spectrum*, 3, 3 (1968), 25–9.

# 10

# Clashing Knowledge-claims in Nineteenth-century English Vaccination

*Logie Barrow*

From the 1850s, smallpox vaccination was among the most 'wilful' acts hitherto required by the British state of all subjects. Indirect tax-paying, its main lifeblood, was no more than marginally voluntary. Unlike many Continental equivalents, the state required no residential registration nor, for males, compulsory military service if we exclude the part-time and far from universal militia and, into the Napoleonic wars, press-ganging near the coast. Other intrusions were partial, such as tithes, or rare, such as jury service. Yet, from 1853, all babies in England or Wales were to be vaccinated, usually by the officially recommended method of arm-to-arm, within three months of birth. This was earlier than in Scotland or Ireland, let alone in most Continental countries. Exactly seven days after the operation, every 'vaccinee' was to be presented for possible extraction of the resultant 'lymph', i.e. vaccine, for vaccinating the next batch: to become, in the jargon, a possible 'vaccinifer'.

The point, here, is not to go over yet again the chronology of vaccinal legislation and resistance with its climax in loosenings of compulsion in 1898 and 1907, nor over the mid and late nineteenth-century increase of state intervention in subjects' lives, nor over the professionalization of medicine, a process often denounced by its opponents as a conspiracy to monopolize, even to pervert, knowledge. Rather, the point is to examine how stunted vaccination's mediation was by pro-vaccinists' disdain for the 'poor' majority of parents. How deeply did the latter lack 'intelligence'? At the time, this concept was deployed more frequently in debates as to when or whether some or all working-class males should be given the right to vote, lest they win it for themselves. Both discourses, vaccinal and political, were ultimately about the human validity of the majority of one's fellow-Britons.[1]

We can at once observe some curious results for the mediation of matters vaccinal. Upholders of vaccination sometimes discussed their problems as if their journals were internal memoranda. 'A death from [blood-poisoning 'from vaccination'] occurred not long ago in my practice', the Medical Officer of Health for Aston, Birmingham, Henry May, confessed during the winter of 1873–4 to his local Medical Society, 'and although I had not vaccinated the child, yet in my desire to preserve vaccination from reproach, I omitted all mention of it in my certificate of death'. As he had reasoned a few lines earlier in his speech, 'it ... [was] scarcely to be expected that a medical man will give opinions which may reflect upon himself in any way'. 'In such doubtful cases [a responsible medicator would] most likely tell the truth, but not the whole truth, and assign some prominent symptom of the disease as the cause of death.'[2]

May would have disagreed with our word 'confess': he surely knew that his more or less routine address would, like others to this particular Society, be printed in the *Birmingham Medical Review*. Thus the wonder is that he behaved as if his words would circulate exclusively among the like-minded and confidential, despite encapsulating what anti-vaccinators (henceforth, 'antis') had long been saying: that many medicators tried to hide the potentially lethal effects of vaccination. 'Antis' greeted May as letting the cat out of the bag; so they made sure that its screeches echoed for decades. They welcomed similarly his expression of trepidation that Parliament might soon try to force medicators to state causes of death 'truly'.[3]

Nor was he unique. Six years earlier, one prominent West Country surgeon had, unlike May, seen himself as a whistleblower. Robert Brudenell Carter, a Fellow of the Royal College of Surgeons, had certainly chosen a far more conspicuous stadium, the *Lancet*, to complain of 'a sort of common consent among medical writers to gloss over the evils that may be attendant on vaccination, for the sake of its great and manifold benefits'.[4] 'Gloss[ing] over' was surely a hard charge to disprove without appearing to exemplify it. Yet the *Lancet*'s editors, like their *Birmingham* counterparts, saw fit to print such accusations. We will view some further own-goals (as they became) from the orthodox later, underlining how peculiarly fraught the vaccinal area was for any 'mediation' from orthodox medicators to any class of laypeople. Worse from any orthodox viewpoint, we shall soon see the 'knowledgeable' receiving, over the decades, as many unwelcome lessons as the 'antis'.

Editors and many contributors to the medical press often behaved as if every 'anti' were illiterate or penniless or both: to be ignored anyway.

The same condescension extended to officially qualified colleagues who had become 'anti': thereby, they had sunk beneath contempt, not least because their formal qualifications made their criticisms all the more welcome to 'antis'.

Thus, medical discussions long remained more monolithic than political ones. On the key political question of the franchise, there were two versions of 'intelligence': monopolist and pluralist. The first took intelligence as good taste: more or less the same as one's own, it would never 'endanger the constitution'. This was not necessarily to deny further franchise extension, in the manner of Robert Lowe or the future Lord Salisbury during 1866–7. True, intellectual monopolism certainly deepened anxieties about extension: Salisbury, premier into 1902, even publicly loathed 1867 as transforming politics into slow-motion ruling-class suicide, with himself a leading beast among the Gadarene swine. But, from Disraeli down, most proponents of franchise extension argued likewise from monopolist assumptions. Yet not all: during the same crisis, John Stuart Mill formulated a pluralist intelligence. 'We all of us,' he had reminded his fellow-MPs,

> know that we hold many erroneous opinions, but we do not know which of our opinions these are, for if we did they would not be our opinions ... Every class knows some things not so well known to other people, and every class has interests ... I claim the benefit of these principles for the working classes.[5]

Contrast, in the vaccinal field, the President of the Local Government Board during 1898. 'Intelligent people,' Henry Chaplin proclaimed, 'desire to promote vaccination ... and the whole ... clamour against it comes ... from want of intelligence.' Here, he was damning any argument for allowing a new right of conscientious objection to vaccination. 'But I want to spread the intelligence', one pro-conscience MP interjected. 'Yes, and so do I', Chaplin snapped back. Other than this instant verbal flick, though, he offered no reply.[6] For him, allowing rights to 'conscience' undermined 'intelligence'; for the MP, it 'spread' it. Mere days later, not argument but by-electoral pressure forced Chaplin to allow such a 'conscience clause' after all. Many of his fellow-Tories were stunned and outraged. Ironically, as we will see, Salisbury would have to help rescue him, partly by explaining to the Lords how not to be trampled.

The internationally feted disappearance of smallpox (barring negligence as in late-1970s Birmingham, and current fears of bio-terrorism) may leave any fresh readers wondering how anyone could ever have opposed vaccination. However, once immunization against anything is even alleged let alone offered, most parents in any century or society are condemned to weigh the dangers and current likelihood of the disease against the risks and effectiveness of the preventative. Obviously the intricacy of such balancings can vary. Worse, nineteenth-century vaccination was, in Baxby's nice anachronism, 'ahead of its time' or rather of 'its' science.[7] Not least, Jenner's promulgation of vaccination at the end of the eighteenth century occurred three generations ahead of the arrival of laboratory methods we would now judge appropriate to safe, swift and flexible production of vaccine. Most orthodox medicators tended to agree with the most meritocratic and democratic of their propagandists, Thomas Wakley (1795–1862), in identifying vaccination as 'the greatest boon that science ever gave mankind'.[8] But they must sometimes have felt like someone who had mounted a penny-farthing without understanding any principles of locomotion: ounces of practice might be worth tons of theory, but only as long as you stayed on top.

This general insecurity was one reason why protagonists on either side tended to emphasize how high the stakes were. Another reason was that compulsion was legislated (1853) at a time when orthodox medicine was still controversially acquiring (with what became the 1858 Medical Act) approximately the structure and powers it has to this day, thus intensifying a situation in which its legislative reach exceeded its curative grasp. Given this potential over-exposure, Wakley himself (for whom 1858 would soon be an incompetent bungle) prophetically warned that compulsion would one day boomerang: it would 'bring vaccination into unmerited and irrecoverable disgrace' and, even worse, prove 'suicidal' for 'the [medical] profession'.[9]

Thus pro-vaccinists descanted on more than the horrors and infectiousness of smallpox, 'antis' on more than the risks and alleged uselessness of vaccination: this propaganda war went beyond encouraging any parent to allow or refuse the operation. Rather, the dynamics of the struggle widened the target medically and politically. There were many dimensions: 'compulsion' and 'freeborn' Englishness; public and private health; central and local; medical orthodoxy and heterodoxy; professional and amateur; elite and popular. These issues (and others) often overlapped, though not always straightforwardly at all. The one I wish to stress most here, is one which has recently (over at least MMR, i.e. combined immunization against measles, mumps and rubella) become obvious again, a century later: impatience among professionals and their political backers undermines itself when outraging the mostly informal conceptual cultures of non-professionals.

To rephrase more pointedly: the greater the tension between more or less democratic and elitist epistemologies, the greater the likelihood that the epistemolgical will sooner or later convulse the political. We have fleetingly glimpsed this occurring in 1898 when popular pressure contributed to what at first seemed a partial repeal of compulsory vaccination. England's vaccination struggles were epistemological in the extreme sense that the two main sides tended to deny each other's ability to think straight. The same struggles may have been the most resoundingly mass epistemological ones in nineteenth-century Europe (outside some cantons in Switzerland, a country with a 'free' self-image as strong as England's). English 'antis' tended to share with the heterodox in many other fields what I call a democratic epistemology, that is, a definition of worthwhile knowledge as comprehensible to anybody and never to be made incomprehensible, or otherwise inaccessible. They enjoyed adopting a pose as defenders of their own and everybody's medical rights against invasive compulsionists. 'A greater mistake was never made,' one Southwark 'anti' thundered to the Liberal *Daily Chronicle* during 1878,

> than that which supposes that all knowledge of medicine ... is wrapped up with the diploma to which is given State patronage, State support, and State license [sic]. There are men without these who can, and do, administer medicine with great success.[10]

Thus, no anxious parents needed any Michel Foucault in order to fear compulsion as disempowering them intimately, however many millions of under-three-month-olds were vaccinated, often in more or less slapdash ways, from 1853 into the 1890s. Now and then, though, the epistemologically and medically mighty were confounded. Some of their oldest conundra appear only now to be moving towards settlement, as over the viral relationship of smallpox to vaccinia, the disease which Jenner had seized on as conferring immunity against it. Were the two related or identical? If the latter, then vaccine was merely the old smallpox virus whose inoculation had become illegal from 1840.[11]

Other controversies had been on the whole hurriedly forgotten, whether settled or not: into the middle of the nineteenth century, many orthodox medics had believed in drawing a little blood while vaccinating (in other contexts they often drew much), whereas their later colleagues increasingly recognized this as dangerous, particularly if blood were mixed with lymph which might then pass into other vaccinees. But what about microscopic quantities?

From here on, we will see historians tending, implicitly at least, to over-emphasize the importance of official prescriptions, great though this was. Years of wading through the medical press gives a more complex picture. For, in fact, disagreement continued over almost any detail of the operation. True, from the late 1850s, official standards were being enforced on public vaccinators, but only chancily, via often brief and perhaps arbitrary visits by an inspector from Whitehall every two years or so. Private operations were virtually uncontrolled. True again, one controversy seemed to be drawing to a close. Did the degree and length of immunity vary with the amount of lymph inserted? Increasingly from mid-century, the answer became 'yes', and the recommended number of insertions (four) was emphasized with mounting insistence. But not always successfully: complaints remained constant from public vaccinators (PVs) that parents tried pressuring them to make as few insertions as possible or simply went to 'sixpenny doctors' who would oblige. For PVs and their supporters, this openness to parental ideas was treason to vaccination and thus to science.

Worse, while most early pro-vaccinists (including at first Jenner) had talked as if the operation conferred total and permanent immunity, complexities had soon become discernible. True, some mid-century specialists still saw any mention of these as the naivest mediation, if not treason again. But, sooner or later, shifts became clear. Nicely, Anne Hardy quotes two consecutive annual reports by the Registrar General. In 1881, he still spoke from the same mental universe where, back in 1857, John Simon (one of the century's key health policy-makers) had hailed vaccination of infants as conferring permanent immunity 'when properly performed' (that fast-growing weasel of a phrase). Yet suddenly, in 1882, it was 'pretty generally recognised ... on good grounds, that the immunity from vaccination is both less perfect and less permanent than that ... [from] smallpox itself'.[12] By the 1890s, the main controversy was not whether to revaccinate, but rather when and how often. Some pro-vaccinists wanted to make at least one revaccination compulsory.

Most emotively of all, there was 'syphilis': one of the most feared diseases, partly because till the 1900s difficult to diagnose. Into 1870, rumours that vaccination might spread it were loudly denied: where these were confirmed from France or Italy, then so much for the sloppy, hysterical or even amoral continentals. But in April of that year the surgeon Jonathan Hutchinson, President of the Hunterian Society, informed an 'unusually large attendance of members and visitors' at London's Royal Medical and Chirurgical Society that he had seen a handful of local cases of vaccinally transmitted syphilis. Though he soon gave most of these up, the cat would not retreat to its bag and became all the more raucous, the more some medical journalists berated Hutchinson for mentioning them at all.[13] No wonder that Simon's immediate reaction, at that same Royal Medical meeting, was virtually to wring his hands in public:

> if every parent was to think that syphilis was likely to be given to his child when vaccinated, such an idea would, if anything could, justify a rebellion against the law. The profession ... must examine into all the sources of danger.[14]

This was a rare public admission from a key activist of orthodoxy, whose evidence during the same weeks to the House of Commons Select Committee on Vaccination was often evasive to the point of arrogance. Noting his outburst via the *British Medical Journal*, 'antis' were happy to quote him in order to encourage, precisely, 'rebellion against the law'.

Our metaphorical cat became yet noisier from 1883, when the government's 'Chief Vaccinator' Dr Robert Cory managed to transmit syphilis, with the vaccine, to none other than himself: for some, the defence, in itself plausible, that he had freely and heroically gone beyond the call of duty by ignoring all the rules was not quite the point.[15] More sweepingly, during 1887-9 two prestigious researchers, Dr Charles Creighton of Cambridge and the (for Britain) pioneering bacteriologist Professor Edgar Crookshank of King's College London, insisted that vaccinia was closely related to syphilis. Thereby, they might license a relabelling of many vaccinal 'transmissions' of syphilis as merely vaccinia reverting to type.[16]

Though Creighton and Crookshank wrecked their careers for what was then widely lambasted as treason to vaccination and can now be recognized as a fallacy, they (plus Cory's self-infection) helped trigger a Royal Commission on Vaccination. This proceeded to sit for an unexpectedly long period (1889-96). Coincidentally during these very years, formal democratic control over local authorities was increasing: in 1888, for example, county councils replaced local rule by Justices of the Peace (JPs, when sitting in quarter-sessions). Even the Poor Law guardians (who, from the start of public vaccination, had been saddled with the task of administering it at the local level) were under pressure directly or atmospherically. By the late 1890s, one-third of the boards of guardians in England and Wales were refusing to enforce. By then, a backlog of unvaccinated young people was a decade-old feature in London where smallpox seemed almost endemic: Hardy notes increases in the proportion of 'infants not finally accounted for' vaccinally in sixteen of the thirty metropolitan Poor Law unions between the late-middle 1870s and the decade of the 1880s; but these proportions became far more serious during the 1890s: in three unions to over 50 per cent.[17]

During the nineteenth century, divisions between democratic and elitist epistemologies tended increasingly to overlap with those between heterodox and orthodox medics,[18] even though the former certainly enjoyed some support from the socially elite and even though some orthodox practitioners were, like Wakley, politically far from supporting any existing elite. Unlike the eighteenth century,[19] the nineteenth saw growing polarization between heterodox and orthodox medicine, not least because the official structures of the consolidating medical profession discouraged the one and encouraged the other. Vaccination became the key struggle in orthodox and official medicators' attempt to assert responsibility over the whole population just as, in the three Contagious Diseases Acts (all passed during the 1860s), they attempted to assert responsibility over a part.

Success against those Acts was to come more swiftly (in the 1880s)[20] than against compulsory vaccination. Unlike syphilis, smallpox was not perceived as morally discriminating; for this and other reasons, there was even more at stake in the struggle against it. But, both nationally and locally, the two 'anti' movements shared many activists and tropes. The latter included outraged purity (female or infantile) and pollution by Act of Parliament. As Durbach most recently instances,[21] both movements used the full available range of propaganda techniques: meetings, marches, elements of ritual humiliation against enforcers and their shibboleths.

The historiography of vaccination is dominated by two in themselves indispensable lines of research, each of which generates a particular tunnel vision. The first emphasizes legislation and the struggles against it. Pioneers here include Beck, Lambert and Macleod. Durbach has recently narrowed the focus to the 'antis', applying greater sympathy and postmodernist alertness to discourses and ideology. Dorothy and Roy Porter retain the interactiveness while concentrating on one major local episode. Others have narrowed the same focus even more to legislation. This has generated some capital international comparisons, not least Hennock's. Above all these, Baldwin's massive scintillation covers a range of diseases within four countries.[22] A second line emphasizes official medicators and their fight against diseases. Anne Hardy is a fine example here.[23]

But, however much some of these authors look at both sides in the vaccination struggle, they tend to elide non-vaccinating parents into 'antis' (an elision which contemporaries often avoided via a condescension of their own: labelling the non-vaccinating as lazy). They also tend to reduce anti-vaccinationism to the more hysterical, often more or less religious, pronouncements by ideologues. Obviously, parents often used religious (and secularist parents no doubt anti-religious) language too. But this usually came after they had weighed the risks. It is this parental weighing of risks that is too little respected. Where epidemic fears helped numbers of previously non- or even anti-vaccinators to have their children vaccinated after all, the Porters' otherwise fine article derides their previous stance as shallow. Were such derision ever relevant at all, was compliance never shallow? No authors attempt more than gesturally the trickier task of reconstructing relationships, i.e. mediation, not least medical, between and among parents, vaccinators and pro-vaccinists. Thus Durbach's concentration on one of these many sides involves over-reliance on 'anti' sources as to how the laws and the operation itself may have felt at the receiving end. Baldwin's concentration on legislators and regulators involves a similar tunnel vision, symbolized in the title of his whole vaccinal chapter: 'Smallpox faces the Lancet' breathtakingly implies that lancets were invariably the implement. Hardy's concentration on medicators tends to elide those they medicated, as in her uncharacteristically sweeping and unfootnoted claim, made in passing, that a particular method of vaccination 'was preferred by parents'.[24]

Urgently required are perspectives on what parents may have thought of which method, and why and when some preferred forgoing any vaccination altogether. We need indications of what medicators (the historiographically obscure at least as much as the memorable) may have thought about the operation and its value. Such thoughts were likely to impregnate their treatment (or sometimes, partly depending on your perspective, mistreatment) of vaccinees and parents. Until we can glean more such indications, we are in danger of treating medicators as condescendingly as they, or many of them, may indeed have been themselves. I used, myself, to relish such counter-condescension.

Coincidentally, during the same generations as a growing minority of men won the chance of recognition as full political adults (mostly via franchise extensions in 1832 and 1867), what many parents of all classes experienced as vaccinal infantilization, compulsory vaccination, was fastened (1853) and tightened (1867 and 1871) on the whole population. During 1866–7 a revealing contrast occurred: Parliament agonized over the second Reform Bill, while imposing further vaccinal compulsion in a no more than slapdash way. Preconditions for the 1890s crisis of compulsory vaccination were: the 1884 third franchise extension (to something like two-thirds of adult males, plus seat redistribution during 1885) and, as we have seen, a partial democratization of local government.

For most poorer people, public vaccination had three disadvantages against private, whether the latter was pricey or sixpenny. First and worst was its administration by the Poor Law guardians. True, publicly funded operations were supposed (or so a one-clause Act had proclaimed in 1841) to be 'non-pauperizing', i.e. not to degrade parents to objects of the 1834 Poor Law. But, apparently with many parents, such a clause seemed mere words: the PVs and other personnel, such as 'Vaccination [enforcement] Officers' (who had to be appointed in every Poor Law union from 1871), were frequently Poor Law employees already, and the general atmosphere perhaps overwhelmingly Poor Law. No wonder that the government's own expert on vaccination law, D. P. Fry, told an 1871 Select Committee of the Commons on Vaccination that the dissociation from the Poor Law was 'a little fanciful'.[25]

Secondly, Whitehall strengthened this association with the Poor Law by forcing PVs increasingly to do their operating at public 'vaccination stations', particularly from the late 1860s. This was because it mistrusted the hygienic quality of 'stored' lymph: such lymph was indeed highly untrustworthy until glycerine began to be properly used. Whitehall, years after its Continental equivalents, first woke up to the full significance of this procedure in 1898. However frequently or infrequently stations might be identical with Poor Law buildings, the Poor Law's salience in working- and lower-middle-class lives was almost bound to make any 'public' grouping of people potentially degrading.

Third, at these stations, with their PVs often stressed all the more by pressures from both Whitehall and the local guardians, parents were often made to feel that their babies were extensions of an arm or arms which might produce usable lymph a week later for the next in the series: battery chickens (we might say), laying vital vaccine.

Meanwhile, though, interaction between epistemology (as defined above) and politics was far broader than merely vaccinal. True, universal male suffrage came only in 1918; full women's suffrage ten years later. But the 1872 Ballot and the 1883 Corrupt Practices Acts (and not merely the enfranchisements in 1867, 1884, 1918 and 1928) ensured that generations were convulsed not only by anticipation, eager or fearful, of formal political democratization[26] but also by disagreements as to the right of the masses to take educational and medical care of themselves. The educational ones included patchy concession of School Boards under the 1870 Education Act, but then the abolition of these in 1902; denigration and wrecking of working-class 'dame' schools; and Social Darwinian warnings about the stupid and fast-breeding masses, warnings that foreshadowed a twentieth-century obsession with 'IQ'. The medical ones included, not merely the repeal of the Contagious Diseases Acts during the 1880s or the loosenings of compulsory vaccination during 1898 and 1907, but also a worried 1910 *Report as to the Practice of Medicine and Surgery by Unqualified Persons.*[27]

Thus, any contemporary who scanned the copious reporting on the debates preceding the 1867 and 1884 franchise extensions might sense and often witness how much they were about the 'intelligence' of, respectively, minorities or a majority of working men. 'Intelligence' denoted acceptance of the rules of the political game. By contrast, the parliamentary discussions which led to the coercive vaccinal legislation of 1853 and 1867 were often ill-attended and occasionally perfunctory. By contrast again, though, the debates of 1898 were long and agonized. These led to what seemed like a partial repeal of compulsion by giving 'the' parent, usually the father, a right to try and convince two magistrates (or one full time 'stipendiary') that he had a 'conscientious objection' to vaccination. 'Conscience' in vaccination functioned rather similarly to 'intelligence' within discussions over the franchise: by claiming it, one risked forfeiting from those above one any creeping recognition that one possessed it. For many Tories and doctors during the debates and for many magistrates afterwards, the mere act of claiming to have a conscience signalled one's stupidity and callousness about one's neighbours' attempts to avoid infection by smallpox.

Hence the agonizings over vaccinal conscience. We shall now see how deeply relevant these were. The summer of 1898 marks indeed a long-forgotten crisis for a Tory government. Among both causes and effects of this forgetting is the fact that, in the event, that government was to last till January 1906. The crisis was mainly over those consciences. Tories are not famous for tenderness to such things except, during the early eighteenth century, their own. By 1898, they were a mere four years off realizing their dream of abolishing the School Boards. At the time, these were not only the most democratic and sometimes, thanks to the electability of women to them, gender-blind of British public authorities. They were also the most conscience-sensitive, thanks to their electoral system which was not only proportional but also allowed any strongly motivated elector to 'plump', i.e. to give multiple votes, for a particular candidate. This mechanism fulfilled its aim of accommodating religious minorities. But all this the 1902 Education Act abolished. It thus triggered widely reported 'conscientious objections' to paying for all or part of publicly funded education. Here, though, there was no legal mechanism for recognizing consciences: not even a bumpy mechanism as, from 1898, over vaccination. So, over education, 'conscientious objection' could become 'passive resistance'. This was to worsen the Tories' landslide 1905–6 defeat. So the Tories' retreat on vaccinal consciences contrasts with their self-damaging outraging of religious ones over education.

In the Lords during summer 1898, Prime Minister Lord Salisbury and, in the Commons, his deputy and ultimate successor Arthur James Balfour, led the government's seeming U-turn over consciences. Unusually for top British politicians in any age, both were intellectual heavyweights. They had to deploy all their lugubrious dialectic to galvanize Tory backbenchers into this uncharacteristic manoeuvre. 'Some gentlemen,' Balfour pleaded,

> would put [the hardening of public opinion against vaccination] down to the fault of the doctors, and undoubtedly there has been some change in medical opinion ... But ... I don't admit that the medical profession and scientific opinion are to blame. After all, if science is anything, science is progressive, and you cannot have progress without modification of [previously] accepted truths.

Nor did he even 'reproach ... the medical profession that they should have stood out for facts with a dogmatic assurance ... [later weakened] by subsequent investigation, but ... which were in the main true, though somewhat overstated'. This hardly made 'the doctors ... infallible. They have made enormous mistakes in the past, and are predestined to make enormous mistakes in the future; but it means that, however ignorant the doctor may be, he knows at least more than the anti-vaccinationist.'[28] In his patronizing way, Balfour was of course swatting 'antis' with a then, as now, very familiar elitist epistemology:

in this world (for him, a vital limitation) no one would ever securely know anything, but doctors' incomplete knowledge was by definition always preferable to anybody else's. Doctors should therefore retain a licence to 'make ... mistakes' with other people's health. Even more cleverly, by the very act of taking 'the doctors' under his wing, Balfour was implying that they, among pro-vaccinists, had kicked all the own-goals. Yet mistakes no less strategic had been made no further from Westminster than Whitehall, where the officials controlling public vaccination had continued insisting on stational arm-to-arm technique till no more than weeks before Balfour rose to his feet, even though on the Continent the method was now widely superseded by laboratory-based methods which some British medicators, too, had long been advocating.[29] This was one reason why all sides bitterly agreed with Lord Lister, the recently ennobled grandee of antisepsis, when he pleaded with his peers for what he called the 'tremendous experiment': evidently, millions of lives and much else were involved during the 1898 switch of public vaccination from stationally based compulsion using arm-to-arm lymph, to the new panacea of glycerinated calf-lymph tempered with an indefinite number of officially unintelligent consciences.[30]

Unlike Balfour, Salisbury and many historians, some 'antis' shared Mill's pluralist definition of intelligence. 'I don't think ... [vaccination] is a case for Parliament to interfere', the Swedenborgian Dr J. J. Garth Wilkinson informed the Commons 1871 Committee. For him, vaccination should, like 'everything connected with the medical art', be left to 'Free Trade'. His grounds were not only strategic but also logical:

> Legislation ... fixes medical dogmas ... I think if Parliament would ... leave us [medicators] alone to rival each other, and to fight these cases out, medicine would be in a much better ... state than it is.[31]

There is no surprise in the medically heterodox supporting what Wilkinson called 'medical freedom'. Clearly, the 1858 Act had not aimed to free up the medical market. Nor, perhaps, should too much be made of support for medical free trade from an occasional orthodox bigwig, such as Benjamin Ward Richardson, a pioneer of anaesthesia and of much else. Significantly, though, Richardson was mildly suspicious of vaccination, deeply so of Jenner. Conversely, a move towards heterodoxy was not necessarily identical with one away from vaccination: Wilkinson had vaccinated occasionally for more than two decades after his shift to homoeopathy. What is striking is that most 'antis', like most of the heterodox in general, viewed compulsion as merely bringing an arrogance, that had always been part of orthodoxy, home to anyone at the point of a state-empowered vaccinal implement. The belief thus followed, that medical free trade would involve not only ending compulsion but also pulling down orthodoxy from its seat and exalting the open-minded and meek. Rhetorically, the two latter categories were often assumed to be near-identical.

The internal transformations of orthodox medicine during the century, involving, for example, a conspicuous retreat from bleeding and inoculation and an embracing of revaccination, made it an easy target for polemic. What further loaded the atmosphere was a widespread lay perception that, not only had 'doctors arrogated to themselves all knowledge', but also (I am quoting from speakers welcoming home to Walsall, near Birmingham, a father-of-two who had gone to prison for non-vaccination during, again, 1876) that 'vaccination ... was not for medical opinion [to decide], [but] for practical observation'.[32]

We now have some context for the (at first sight) cheekiness with which one of those who had led in making Leicester into one of England's most anti-vaccinal towns stood up to one of Britain's leading venereologist-surgeons. 'It is simply,' Jonathan Hutchinson asked during 1891 as a member of the Royal Commission, 'the belief of the father of the child ... There is no medical opinion in that?' 'I am unable to say', J. T. Biggs replied, 'what medical opinion the father might have had.'[33] Hutchinson did not take up Biggs's quiet insistence that a mere father could *have* any 'medical opinion' that could ever stand comparison with the opinions of those medically qualified to have 'opinions'.

One aspect of professionalization was professional capture of a few popular remedies. Vaccination itself had begun as a classic example. As a committee of MPs had admitted during 1802 (in conclusions reproduced by another committee of MPs during 1871), some of those 'common people employed in dairies' in 'various parts of England' who had noticed a correlation between cowpox and immunity to smallpox had probably 'in some very rare instances' carried their knowledge 'one step further' in order to spread cowpox 'on purpose'.[34] Thereby, such 'common people' would have been immunizing themselves and perhaps others against smallpox without intervention by professionals. We can doubt whether legislators could have been so respectful by late-century, despite being formally subject to greater democratic control: one of the epistemological paradoxes associated with Britain's simultaneous professionalization and partial democratization.

Wilkinson and Richardson, the two surgeons whom we have just met, were very rare in recognizing possibly illiterate local women as colleagues or legitimate competitors, to whom many parents turned as a first line of medical succour. But the general official disdain was, in vaccination, self-defeatingly one-sided. Not that one-sidedness was an official monopoly: some 'anti' journalists reduced the whole of vaccination to a money-grubbing conspiracy,[35] in effect homologous to the old Enlightenment reduction of religion to 'priestcraft'. Nor were all 'antis' invariably proof against snobbery: one 1878 propagandist could sneer at 'Jenner's clodhopping superstition'.[36] But not all were so crude. When a member of the 1871 Committee tried patronizing Wilkinson for trusting the uneducated more than the educated, he again echoed Mill's pluralism of five years earlier:

> if there be a class of society in which vaccination does more mischief than in another class, the ... [former] will know more about one particular side of the subject ... vaccination among the poor ... is a very different thing in its indiscriminateness, and necessarily so when vast masses have to be vaccinated, than vaccination among the class in which a medical man waits two or three months to select a child to vaccinate from.

This was obviously because 'indiscriminate vaccination from poor little miserable wretches to other poor little miserable wretches is more likely to cause vile results'.[37] His was the inverse of Whitehall's perspective: for Whitehall, public vaccination was *superior* to private, because under its control.

Some pro-vaccinists would, from their side, have agreed with Wilkinson that vaccination's very medical prestige was stultifying intellectually as well as medically. 'At the Medical and Chirurgical Society the other night,' Dr Robert Hall Bakewell noted during 1871, back from some years as Trinidad's top medical official and vaccinator,

> several speakers praised Mr Hutchinson [our leading venereologist] for his 'courage' ... [in revealing his cases of apparently vaccinal syphilis]. Why should it require more 'courage' to report such cases than to report a death from chloroform, or from pyaemia after an operation? But it does, and it could only be done by a man occupying Mr Hutchinson's high position in the Profession.[38]

Conservatism, though, was an accusation thrown by each side at the other. 'In all sciences,' Wilkinson told the 1871 Committee, 'the

difficulty has been, not to keep things as they are, but to change them.'
This was all the more so in vaccination which suffered from 'a flood of
orthodoxy' and from 'powerful interests which prevent most medical
men from ever investigating the subject. I think 99 out of 100 of my
brethren have never looked at the other side of the question at all.'
Were they to, he believed they would be in trouble.[39]

But most 'antis' were to show a conservatism of their own in their
reactions to repeated orthodox rejoicings over Pasteur. In this way they
were not alone, but too often they made themselves conspicuous in
defence of ageing paradigms. From their side, most upholders of pro-
fessionalism learned to glory in what they saw as the progressiveness of
their science. 'It is no detraction from George Stephenson,' a *Lancet*
writer announced during 1885,

> that the locomotive which now takes us to Newcastle in 6 hours is a
> more perfect engine than ... [his *Rocket*]. Equally, it does not lessen
> the fame of Jenner that time shows two vaccinations to be better
> than one, as future time may possibly show that three may be better
> than two.[40]

This was irrefutable but, with locomotives, efficiency had increased
whereas, the admission now ran, with vaccination there had been
exaggeration. In an earlier generation, Wakley's *Lancet* had hailed
Harvey and Jenner as the chief heralds of what it saw as scientific medicine. This had been plausible even then. But many medicators knew
they were on our 'penny-farthing', where practice wobbled on a basis
of unstable theory. Some pro-vaccinists saw all this as a non-problem:
'Perhaps the scientist or [sic] chemist could explain – I know not – ',
COMMON SENSE protested from Gloucester, 'how and why the vaccine
lymph produces its effects; it is sufficient for practical purposes that it
does.'[41] Others, too, liked to hail current trailblazers without waiting
for much detail. In 1880, T. H. Huxley prominently injected Pasteur
into vaccination debates. He spoke as part of a prestigious medical dep-
utation to the Local Government Board to stop an incoming Liberal
administration giving some flexibility to compulsion. 'The present
moment,' Huxley pleaded, was

> particularly inopportune, from a purely scientific point of view, for
> any action ... which might be construed by ignorant persons into
> an admission that the evidence ... [for] vaccination is in any way ...
> weaker than ... formerly.

This was because 'a very eminent French investigator, M. Pasteur' had just concluded some experiments on 'chicken cholera' which, Huxley repeatedly promised, gave 'by analogy ... for the first time ... something like complete scientific evidence of the value of ... vaccination'.[42] At a strategic moment, reprieved French laboratory chickens outfaced English consciences, in the event for another eighteen years. The Board's President would obviously have agreed with the *British Medical Journal*'s 1883 principle that 'men and their opinions must be weighed, not counted' and that, in vaccination, 'mere numbers [were] of little worth'.[43] More fundamentally – at least for a *Lancet* writer two years later – 'honest people' were anyway 'not accustomed to weigh evidence'.[44] Thus, for pro-vaccinists, the stupid population, particularly from the lower classes, needed vaccinally prodding towards health. We have seen that this was indeed happening.

But whether parents were becoming reconciled to vaccinations is a far more complex question. 'The days have gone by,' one northern editorialist generalized historically, 'when a man [sic] [could] legally be prosecuted for his views either by Churchman or Dissenter. [But now in their] place, the ... bodies and health of our children are not our own.'[45] Again, who needed a Foucault?

If pro-vaccinists rejected a democratic public, they sometimes suspected any fickle and indefinable newspaper-reading one no less: we saw near the start how medicators sometimes treated their periodicals as internal bulletins. Consistently, they tended to damn any airing in the 'lay' media of what they viewed as their internal problems. 'The whole correspondence has shown, if that was necessary,' one writer in the *British Medical Journal* grumbled about an 1886 controversy in the *Glasgow Herald* as to how many post-vaccinal 'vesicles' were needed for full immunity against smallpox, 'how undesirable it is to discuss medical questions ... in the columns of an ordinary newspaper.' This was because the, for the writer, stupid public would rush to that controversialist who advocated the lowest number of punctures. Intentionally or not, that controversialist was thus advertising his practice: one of the hallmarks of a 'quack'.[46] In medical questions, it was as if readers were expected to structure their purchasing of newspapers analogously to the way they structured their railway-ticketing: into first-, second- and utterly third-class medical 'intelligences'.

Vaccinators and pro-vaccinists were thus claiming to work or bring wonders (more or less) which they did not themselves understand, while denying medical understanding to non-professionals of any class. They had thus walked into a tight corner for 'mediating' or diffusing anything. What, instead, was mediated to parents was all too often an assumption that the egg (i.e. the vaccine) was far more important than the chicken (i.e. the vaccinee and potential vaccinifer). Of course, there were always exceptions; but this priority was structured into the laws and Whitehall regulations, which PVs might ignore at peril of purse or even of post. Apart from this, the main message available to parents was that, if anything seemed to go wrong after the operation, the fault was theirs: they were either deceived, careless, syphilitic or living in an insanitary neighbourhood. Yet, not till the 1890s at the earliest can we imagine many PVs practising consistent antisepsis.

The mediations are even more interesting in the other direction – i.e. from parents to vaccinists. In the long run, the latter had by the end of the century to admit that parents might occasionally have been correct in attributing subsequent infection to operations or operators. No less damagingly, all too many vaccinators and pro-vaccinists had retreated far too slowly in admitting the need for re-vaccination. As for Whitehall, its own retreat from stational arm-to-arm methods was particularly belated and outwardly sudden.

Otherwise, we have also seen how Whitehall and many official and orthodox medicators saw the very existence of an 'anti' movement as proving how intellectual independence might bring forth stupidities so crass as to endanger everybody's life. When the 1898 Act triggered large queues for exemption, many experts, from the government's Medical Officer down, warned of cataclysmically educative epidemics. These never materialized. 'Antis' on their side agreed with most (though not all) heterodox medicators in viewing compulsion as the crowning instance of state backing for orthodox mystification and perhaps jobbery: if the Contagious Diseases Acts were parliamentary rape of women (directly or potentially of working-class ones, symbolically of all), the Vaccination Acts were 'parliamentary poisoning' of children: all, literally. Particularly in this perspective, what mediation could be more horrific?

Ultimately and with the greatest wisdom after the event, the story of nineteenth-century English vaccination highlights some paradoxes of formal, though considerable, political democratization accompanied by widespread (and often self-confirming?) assumptions that the widening 'demos' was mostly, even congenitally, stupid: a classic misfit between the political and the epistemological.

Such or similar paradoxes may still be very much alive. All are far from being peculiarly English or even British. Yet through what clearer prism can we view them than via England's political culture? Its often slapdash elitism eased the long-term own-goal that was Victorian public vaccination. So did its secrecy reflexes, which have more recently helped hothouse bovine spongiform encephalopathy (BSE) and its human variant into a worldwide[47] crisis. Its centralism cut many of its medical teeth on nineteenth-century vaccination (as many contemporaries warned); centralistic actionism, cliquishly grounded on GIGO epidemiology, immeasurably aggravated the 2001 (or rather, perhaps 2000–1) foot-and-mouth (FMD) epidemic. Admittedly, the enquiry into BSE has brought forth a mountainous Report, yet with mostly mouselike conclusions: were those contemporaries on every side who found the 1896 *Report* of the Royal Commission on Vaccination unwieldy able to see the sixteen-volume 2002 Taylor Report, they would relativize that adjective. True again, there are no less than three official enquiries into FMD. Together, however, their terms of reference manage to dodge key empirical issues of governance. Undeniably, today's science and medicine are mediated far more professionally than in the nineteenth century. But mediation takes more than media, partly because it occurs within contexts that are political, epistemological and much else. In government and elsewhere, England's particular bias against at least lay publics seems hydra-like.

## Notes

1. For broader ideological interactions, see particularly Catherine Hall, Keith McClelland and Jane Rendall, *Defining the Victorian Nation: Class, Race, Gender and the Reform Act of 1867* (Cambridge: Cambridge University Press 2000).
2. *Birmingham Medical Review*, 3, January 1874, 34–5.
3. To judge from (inadequate) *Medical Directory* details, May's subsequent career seems not to have suffered from this frankness of his. Some echoes: H. N. Mozley of (or from) King's College Cambridge, *Gravesend Reporter*, 18 August 1888, n.p.; via another 'anti': *Smethwick Telephone*, 19 December 1891, n. p.; J. T. Biggs (for whom see below) during the same year to the Royal Commission on Vaccination: 1891, Question 13835 (henceforth given as: 1891 RC, 13835).
4. *Lancet*, 1868, 1, 767.
5. *Hansard*, Commons, 13 April 1866, vol. 182, col. 1259.
6. *Hansard*, Commons, 19 July 1898, vol. 62, col. 332.
7. Derrick Baxby, *Smallpox Vaccine: Ahead of its Time*, Berkeley, Gloucestershire: Jenner Museum, 2001: many thanks to Dr Baxby for an unsolicited copy.
8. *Lancet*, 1854, 2, 35. Wakley's direct authorship of particular *Lancet* editorials is not always certain, but his influence on their content and style usually stands out.

9. L. Barrow, 'Why Were Most Medical Heretics at their Most Confident Around the 1840s? (The Other Side Of Victorian Medicine)', in R. French and A. Wear (eds), *British Medicine in an Age of Reform* (London: Routledge, 1991), pp. 176–8.

10. 'Mr R. Quarrington of Park-street, Borough Market', *Daily Chronicle*, 2 January 1878, n.p. original spelling.

11. Peter E. Razzell, *Edward Jenner's Cowpox Vaccine: the History of a Medical Myth*, and same: *The Conquest of Smallpox* (both Firle, Sussex: Caliban, 1977); Derrick Baxby, *Jenner's Smallpox Vaccine: the Riddle of Vaccinia Virus and its Origin* (London: Heinemann, 1981). My feather in the scales goes to Baxby.

12. Anne Hardy, *The Epidemic Streets: Infectious Disease and the Rise of Preventive Medicine, 1856–1900* (Oxford: Oxford University Press, 1993), pp. 116–17. The fullest (if sometimes uncritical) source on Simon remains Royston Lambert, *Sir John Lambert, 1816–1904 and English Social Administration* (London: Macgibbon and Kee, 1963).

13. *B[ritish] M[edical] J[ournal]* (1871), 1, 448–9, 453, 504–5, 584; *Lancet* (1871) 1, 52, 618, 654–5, 664, 689; *M[edical] T[imes and] G[azette]* (1871), 1, 13.

14. *BMJ*, 1871, 1, 448–9, 453.

15. *BMJ*, 1883, 1, 1187; 1884, 1, 1053.

16. Charles Creighton, *Natural History of Cowpox and Vaccinal Syphilis*, 1887; same: 'Vaccination' in *The Encyclopaedia Britannica*, 1888; idem, *Jenner and Vaccination*, 1889 (an assassinatory biography); Edgar Crookshank, *History and Pathology of Vaccination*, 1889, two volumes (all London).

17. Hardy, *Epidemic Streets*, p. 127.

18. For sketches of much of this, see L. Barrow, *Independent Spirits: Spiritualism and English Plebeians 1850–1910* (London: Routledge, 1986), pp. 146–94.

19. Roy Porter, 'Before the Fringe: "Quackery" and the 18th-century Medical Market', in Roger Cooter (ed.), *Studies in the History of Alternative Medicine* (London: Macmillan, 1988), pp. 1–27.

20. See particularly Judith Walkowitz, *Prostitution and Victorian Society: Women, Class and the State* (Cambridge: Cambridge University Press, 1980); Paul McHugh, *Prostitution and Victorian Social Reform* (London: Croom Helm, 1980).

21. Nadja Durbach, '"They Might As Well Brand Us": Working-Class Resistance to Compulsory Vaccination in Victorian England', *Social History of Medicine*, 13 (2000) 45–62.

22. Ann Beck, 'Issues in the Anti-Vaccination Movement in England', *Medical History*, 4 (1960) 310–12; R. S. Lambert, 'A Victorian NHS: State Vaccination in England, 1855–71', *Historical Journal*, 5 (1962), 1–18; Roy M. Macleod, 'The Frustration of State Medicine, 1880–99', *Medical History*, 11 (1967), 15–40; idem, 'Law, Medicine and Public Opinion: the Resistance to Compulsory Health Legislation, 1870–1907', *Public Law*, 1 (1967), 107–28, 189–211; Durbach, 'They Might As Well ...'; D. and R. Porter, 'The Politics of Prevention: Anti-Vaccinationism and Public Health in 19th-century England', *Medical History*, 32 (1988), 231–52; E. P. Hennock, 'Vaccination Policy against Smallpox, 1835–1914: a Comparison of England with Prussia and Imperial Germany', *Social History of Medicine*, 11 (1998), 49–71; Peter Baldwin, *Contagion and the State in Europe 1830–1930* (Cambridge: Cambridge University Press, 1999).

23. Hardy, *Epidemic Streets*.
24. Hardy, *Epidemic Streets*, p. 118.
25. 1871 Select Committee, Questions, 3973–7.
26. Some influence from James Vernon, *Politics and the People: a Study in English Political Culture, c.1815–67* (Cambridge: Cambridge University Press, 1993), may be traceable via this phrase.
27. *Report as to the Practice of Medicine and Surgery by Unqualified Persons in the United Kingdom* (London: HMSO,1910), preface, dated 1907, by Sir Almeric Fitzroy of the Privy Council Office.
28. *Hansard*, Commons, 19 July 1898, vol. 62, col. 405–7.
29. For all this and much else on vaccination, I can only refer to forthcoming publications.
30. *Hansard*, Lords, 4 August 1898, vol. 64, col. 37.
31. 1871 S.C., 1306, 1407–8, 1440–1. For Wilkinson, see L. Barrow, 'An Imponderable Liberator: J. J. Garth Wilkinson', in Cooter, *Studies*, pp. 89–117 and idem, entry in *New DNB* (Oxford: Oxford University Press, 2004), forthcoming.
32. *Walsall Observer*, 20 May 1876, 3.
33. 1891 RC, 16,085–6.
34. 1871 SC, *Report*, Appendix 3, 390.
35. E.g., *Daylight* [Norwich], 1 October 1898, 9.
36. H. Clarke, *Yorkshire Independent*, 23 August 1878, n.p.
37. 1871 SC, 1548–9.
38. *MTG*, 1871, 1, 526.
39. 1871 SC, 1426–7, 1327.
40. *Lancet*, 1885, 1, 484.
41. *The Citizen* [Gloucester], 28 January 1881, n.p.
42. *Lancet*, 1880, 2, 230.
43. *BMJ*, 1883, 1, 231.
44. *Lancet*, 1885, 1, 393.
45. *Workington Free Press*, 10 December 1881, n.p.
46. *BMJ*, 1886, 1, 994.
47. See, e.g., editorial: 'The Madness Spreads', *New Scientist*, 10 February 2001, 3.

# 11

# Sex and the Doctors: the Medicalization of Sexuality as a Two-way Process in Early to Mid-Twentieth-century Britain

*Hera Cook*

Historians of sexology, the scientific study of sexuality, argue that from the 1890s sex researchers and activists formed a symbiosis with medicine, bringing sex under the new medical gaze and making sexology respectable.[1] They suggest that the medicalization of sexuality was an expression of the power that medicine, the state or patriarchy, supposedly had or perhaps an element of control that hegemonic forces were extending. Sexological change is said to have been imposed from above (on women or homosexuals) and resisted from below.[2] There is little evidence of this in Britain. Sexual change came primarily from below, as a result of the responses of vast numbers of individuals to common circumstances and experiences. The normalization of sexological ideas and the imposition of them upon some resistant groups followed rather than preceded the growth of desire for medicalized sexual knowledge. And a visitor to Britain in 1946 found that, far from lauding the scientific model of sexuality, 'Doctors practising in sex are still unwilling to call themselves sexologists.'[3]

In the early 1920s, medical or other legally available knowledge regarding sexuality had to be obtained from marginalized individuals and sites. The high sales figures for sex manuals published during the interwar period reveal the strong desire to obtain sexual information.[4] Many of the manuals' authors state that very large numbers of people wrote to them for advice and it is unfortunate that the many letters sent to Marie Stopes appear to be the only set that survived.[5] In the 1920s, the manuals were relatively expensive and hard to find. This had begun to change by the 1930s but it is probable that most readers were middle class until the late 1940s by which time wages had risen. Sales of new editions of the manuals published in the interwar period and of new manuals rose astronomically. From the 1940s, and epitomized by the extensive British press coverage given to the Kinsey reports, credence was increasingly given to sex research and psychoanalysis.[6] It is only from the 1950s that the notion that manuals or sexology were the source of new norms is even faintly plausible.

Sexual knowledge came to be mediated by medicine in the interwar period because doctors were able to supply a particular type of knowledge of the heterosexual body relevant to the social and economic changes that were taking place. The experience of the body altered and new discourses of sexuality were built upon that shifting foundation not vice versa. It is useful to begin asking what medicine might have offered to people who were trying to manage, or to take pleasure in, their sexuality in this particular period and culture. A closer examination of the manuals and their authors provides material to answer these questions.

From 1918 to 1941, six medically trained British authors who published sexual advice manuals under their own names and two pseudonymous authors have been found.[7] The impact of medical training and authority will be revealed by contrasting these doctors with British manual authors who were not medically trained in the same period, a group that included four professional writers and one scientist/writer.[8] These authors published their first manual in the interwar period and were shaped by that period, though several of them continued writing well into the 1960s. The chapter is based on the large amount of published material by these authors, which has not previously been examined systematically, rather than on unpublished sources. Over twenty-five foreign authors published sex manuals in Britain during this period but only one had substantial impact: Theodoor van de Velde, a Dutch gynaecologist, who published *Ideal Marriage* (1928). This was expensive and initially difficult to obtain but the wealth of precise physical detail it contained had a great influence on other authors and it also sold to the public in large numbers once it became widely available from the 1940s.[9]

The content of the manuals falls into three main categories. First, information about the physical body and what sexual intercourse is and how it takes place; then second, information about contraception; and, third, discussion of how, when, and accompanied by what emotion, sexual activity should take place. The basic information about the sexual body and sexual intercourse in the manuals has largely been ignored or read as part of a normalizing discourse constructed to emphasize the naturalness of heterosexuality and to reinforce gender roles.[10] Research into sex manuals began in 1967 with

American sociologists, Lionel Lewis and Dennis Brisset, who concluded the manuals presented sex as hard work. In 1970, Michael Gordon and Penelope Shankweiler added that they promoted prescriptive gender roles.[11] Historians John D'Emilio and Estelle Freedman describe the manuals as providing 'elaborate instructions'.[12] Lesley Hall also glosses over the actual content of the instructions, suggesting that the manuals worked by opening up discussion between couples. This implies people merely needed to communicate rather than to acquire new information.[13] By the 1960s, and particularly in the American context, these judgements of sex manuals were correct. But the argument that sex manuals provided needed information about physical practice in earlier decades has been dismissed without serious consideration by historians. Relevant to this is the assertion that *marital* sexuality was acceptable. In late nineteenth-century Britain marital sexuality had been highly constrained; only relatively infrequent and potentially reproductive coitus was acceptable and the 'elaborate instructions' were introducing many couples to new physical practices, not enabling them to make conversation about self-evident activities.

The British authors of sex manuals were part of British society and had suffered, as they saw it, from societal sexual mores just as other members of the society. Their departure point was that many, if not most, people suffered from profound ignorance regarding sexual practice and that this needed to be remedied. Even coitus had been regarded as needing to be controlled and still was by conservative authors. In 1923, Dr Isabel Hutton explained that unrestricted intercourse was acceptable on the honeymoon but not, of course, on an ongoing basis.[14] Discussions of the appropriate frequency of intercourse occur in all the early manuals, usually resolved by explaining this decision is up to the individuals concerned, but giving a range tilted towards the lower frequencies (the authors were concerned to reject abstinence as a method of birth control). Alternative sexual practices were well beyond the ken of most British people. In 1939, Dr Kenneth Walker placed oral sex in a list labelled 'Common Sex Deviations', which also includes coprophagia, or eating faeces.[15] Even in the 1980s, oral and anal sex remained unacceptable to a substantial majority of British women and a somewhat smaller percentage of British men.[16] The emphasis on 'practices' in the historiography on the manuals is misleading; it was sexuality itself that had been proscribed in the early twentieth century, except where it was necessary for another purpose: reproductive coitus. The manual authors, even the most conservative, were responding to and rejecting this attitude.

There were couples who had disregarded these prohibitions just as many people who desired same-sex partners disregarded the prohibition against their activity. Nonetheless, in the 1920s, the diagrams of the body and the elaborate instructions supplied information that many individuals and couples did not have and wanted, as well as offering support for more positive attitudes.

The existence of profound ignorance about sexuality is supported by all British sources of evidence regarding sexual behaviour around the turn of the twentieth century, as is the belief that people should not speak of sexuality.[17] Sigmund Freud initially argued that those who claimed to be ignorant were in a state of denial, and this position was reformulated by Michel Foucault. Ludmilla Jordanova has commented that 'it could be argued that considerable effort has had to be expended on correcting the more common "truths" [Foucault] put into the public domain, especially about the history of hospitals'.[18] So, too, do historians of sexuality need to correct Foucault's 'truths' about late nineteenth-century and early twentieth-century sexuality. Foucault's achievement was to enable the linking of sexuality with political and cultural processes, thus redirecting attention away from the sexual behaviour of individuals and on to society and the construction of sexual categories. However, he reversed Freud's 'repressive hypothesis', presenting a counter-argument that the Victorian period incited sexuality, producing a multiplicity of discourses and the privileging of sexuality as the core of identity. He prioritized a supposed explosion of rarefied and inaccessible professional discourses and a culture of confession reinforced by the Catholic Church and psychoanalysis. Neither of these institutions found any great favour in Britain, nor has it been shown that discourse on sexuality increased to a greater extent than discourse on other topics.[19] Little attention has been given to the limits that even Foucault himself admitted existed in the Victorian period: 'where and when it was not possible to talk about things became much more strictly defined; in which circumstances, among which speakers and within which social relationships ... This almost certainly constituted a whole restrictive economy'.[20]

During the late Victorian period and the first decades of the twentieth century the vast majority of British people restricted their own physical and emotional expression. They strove to shelter children and young people from sexual knowledge and to protect their innocence by preventing them from encountering sexual knowledge. The physical sexual practices through which heterosexual sexuality or gay and lesbian sexuality is expressed are not natural or instinctive but behaviours that need to be learned. Heterosexual practices appear ubiquitous, and homosexual practices natural to homosexuals, only because in most societies and in most periods the circumstances exist that ensure such learning can take place in infancy and childhood.[21] Sexuality is not primarily discursively produced and nor does language provide the boundaries of sexual experience. Much sexual knowledge will never be verbally articulated by the individual; it is discourse only in the very widest sense of the term, which includes patterns and habits of movement, sound and touch that are not accessible to the adult consciousness. It is primarily through the body that sexual knowledge is produced and it is the body that provides the boundaries of sexual experience.

In the late Victorian and the Edwardian period many children, including those who wrote manuals as adults, matured without these physical learning experiences. They had no positive images of adult sexuality and the intensifying prohibition against masturbation prevented many from accepting pleasure in their own bodies. The historian's knowledge of these prohibitions comes through discourse but the male baby learned that touching his genitals except to urinate and the female baby that touching hers for any reason at all was unacceptable well before they learned to speak.[22] Contemporary observers believed that there was a '*spontaneous* feeling of shame experienced by most masturbators' (my italics).[23] The father of the author and doctor, Edward F. Griffith (born 1895) died when he was young and his mother brought him up. He wrote 'the whole family exuded goodness, and religion glowed in every house I lived in and every house I visited'. This, he says, was 'a good way' but 'I unfortunately, had too much of it.'[24] He did not see boys without their clothes on until he was ten, and he was fourteen before he peeped through a keyhole and saw the back of a girl washing. He wrote that 'for years I had suffered from considerable curiosity about the human body ... [but] everything was secretive and curiosity was unsatisfied'.[25] Griffith's biography, from which these quotes are taken, was written towards the end of his life and is clearly mediated by a lifetime of sexual writings and activism. We could consider that these memories are tropes designed to support his self-image as a fighter in the noble cause of sexual liberation and enlightenment. It could be argued that he is redefining sexual harassment and early homosexual desires as acceptable curiosity about the body and normal male heterosexuality. This would suggest he is creating a construction of sexuality that serves hegemonic interests: those of patriarchy and capitalism. This approach will not help us to comprehend the sexual mores of the period in which Griffith is writing. Late and partial awareness of sexuality meant that puberty was experienced as frightening.[26] Many people never married and, as there was a late average age at marriage, puberty was frequently followed by a long wait before any actual sexual experience even for the individuals who did marry. Like many people, Griffith found his desires incomprehensible and worrying.

Griffith, and the other four doctors who began writing before the mid-1930s and for whom there is personal information, all described themselves as suffering from varying degrees of sexual ignorance, similar to that of their patients. Their writing reveals an earnest desire to dispel this ignorance. Griffith emphasized the extent to which he shared his patients' experience:

> When I qualified my knowledge of sex affairs was nil; I had been taught nothing. I soon found that a large proportion of my patients were equally bewildered and came to me for help.[27]

The equally conventional Isabel Hutton (born 1887) qualified as a doctor and went on to have a relatively illustrious career as a psychiatrist. Her manual, which was published in 1923, seems to have been her one excursion into writing or researching sexuality:

> I began to write in the evenings and gradually evolved something that stemmed from my own past ignorance and difficulties and the questions that patients had asked me throughout the years. [28]

The contents of Hutton's manual suggest that her husband's difficulties were as important as her own problems in motivating her to write. Helena Wright (born 1887) presented herself as considerably more confident and with higher expectations of sexual activity when she married at the age of thirty in 1917. Her father was an immigrant from Poland who began his marriage by informing his wife not to expect fidelity and then took a mistress every year for a year. Helena's unconventional sexual mores probably had their roots in an extension of such upper-class male behaviour rather than in the social movements from which female sexual radicalism is seen as emerging.[29] She took several lovers over the course of her life, though she was also an active Christian. She was motivated to write a manual by the realization that her own experience of lack of sexual pleasure was shared by around 50 per cent of her patients:

> First intercourse wasn't painful, but everything felt dead. I didn't want to wound Peter [my husband], but I thought to myself, 'There must be some other way of doing this.'[30]

Similarly to Wright, Kenneth Walker (born 1882) did not present himself as wholly ignorant of physical sexual information but as lacking the information he needed to understand and accept his own body. He was in his late twenties and had completed his training as a doctor in London before he finally had sexual intercourse:

> A medical friend who was in training to be a psychotherapist [suggested I] 'substitute an actual sexual experience for futile imagination concerning it'. At long last but not until several years had past, I took my friend's advice [and visited a brothel].[31]

However, Walker implies that he, and many other young men, found visiting a prostitute was a problematic route to sexual knowledge:

> During my own boyhood in the Victorian age, sex was regarded as being a guilty secret, a degrading activity, a low animal instinct ... When during their growth signs of this shameful and mysterious force began to become evident in boys, they were at first frightened. Subsequently these signs were accepted as part of a guilty but not unpleasant secret. After growing up the boy in question might marry and find himself impotent because sex for him was still a sin and a shameful secret that had no relation at all to his love for his wife.[32]

This route to sexual knowledge would have been unacceptable to Griffith, but his attitude to his sexuality was similar: 'I had firmly decided by age fifteen that I was going to marry because I did not see how else I could deal with the awful urges inside me.'[33] Eustace Chesser (born 1902) was brought up in a disciplined Jewish household, the son of people who had fled from Russian pogroms. The writer Paul Ferris, who interviewed him in 1965, wrote that his background and public school left him an outsider. His manuals reinforce the impression that he was a less conventional and earnest man than the other doctors, but he still saw dispelling ignorance as important.[34] The authors explain that their encounters with patients convinced them that their own difficulties were shared and even surpassed by a large proportion of people within the population. They then began to write in order to share their solutions. All of these doctors were not merely mediating medicine but mediating their experience as individuals within the sexual culture into medicine. In spite of the insistence by historians that the creation of norms was one of the main products of sexology, the best-selling medical manual authors of this early generation do not emphasize norms but rather reassure readers that variation is itself the norm. They are uncomfortable with many activities but they struggle to be accepting and the only variation they absolutely reject is cruelty or sex that is not mutually desired. However, the historian Margaret Jackson has argued that sex manuals of this period were 'extremely didactic and prescriptive'.[35] According to her, the aim of the authors was the 'securing [of] female consent to male dominance and female submission by eroticizing it and defining it as "natural"'.[36] The claim that they assumed men were responsible for initiating sexual activity and that women needed and wanted to be 'wooed' is correct.[37] But there is no evidence that an alternative and prior construction of het-erosexual physical sexual activity existed. The manuals should be read as revealing that gender roles in physical sexual practice were rigidly constructed and the authors were, in fact, attempting to expand the range of acceptable sexual behaviour. *Married Love* (1918) by Marie Stopes, originally a coal scientist, was the most influential of all the twentieth-century manuals. Her insistence that women should deter-mine the timing of intercourse according to their own sexual desire was feminist and radical.[38] The abuse that succeeding generations of men have heaped upon Stopes makes it obvious how unacceptable it was for women to argue directly that they should play a more active role in sexual activity.[39] The demand made by Stopes and other writers that the woman must be courted (or persuaded, seduced, wooed etc) prior to sexual intercourse contributed to the erosion of male conjugal rights. They were arguing against women being treated as permanently sexually available once marriage had taken place, and insisting that their desires must be taken into account.[40]

Few historians of sexuality have focused on the contribution made by birth control to sexual change. Jackson does do so, but unfortu-nately she derives her analysis from an argument put by some first-wave feminists 'against man's unrestrained exercise of his passions' and in favour of abstinence from coitus.[41] Astonishingly, Jackson argues that the presence of birth control 'makes it impossible to use the fear of pregnancy as a reason for refusing male sexual demands for coitus, and thus makes it more difficult for many women to resist those demands, and to define and control their own sexuality'.[42] But a study of women with *large* families by Pauline Shapiro revealed that 'the husbands shrugged their shoulders [and reflected that children] were not their business'.[43] The absence of contraception does not give a woman the power to define her own sexuality or even to prevent her husband from insisting on intercourse.

In Stopes' *Married Love* she defended contraception but she did not tell her readers about methods until a few months later when *Wise Parenthood* (1918), the first modern manual on contraception was published.[44] Manuals published after *Married Love* not only defended the use of contraception but also provided substantial information about methods.[45] At the beginning of the 1920s there was very limited knowledge as to which birth control methods were effective and why. Even in the 1930s contraception usually remained a matter of self-help in the face of advice that was often confusing. By 1947, Edward Griffith, by then one of the founders of the marriage guidance movement and the author of *Modern Marriage and Birth Control* (1935) could state that '[o]ur knowledge of this subject has advanced so much in the past five years that one is able to state the position much more simply, and in a sense dogmatically'.[46] This knowledge was based upon scientific findings about women's fertility cycle, the collection of statistics on use of methods and improvements in rubber technology. All knowledge is socially constructed but not all socially constructed knowledge will serve the same ends. For example, however strong a woman's belief in sitting up in bed and coughing violently as a birth control method, it will not lessen her risk of conception.[47] The increase and systematization of knowledge about *effective* methods of birth control, which took place during these decades, was an innovation of great importance in the history of sexuality. Kenneth Walker wrote in 1940 that:

> Whilst methods of limiting childbirth are not entirely new, it is only within the last thirty years that contraceptives have been made reliable and brought within reach of the masses. This ... has allowed of the sexual relationship being treated quite differently from how it was treated two or three generations ago.[48]

It is no coincidence that he writes about the 'increased freedom of women' on the same page. From the last quarter of the nineteenth century, men and women had increasingly lowered their fertility but initially this had been possible only by restricting the frequency of sexual intercourse.[49] It was as a result of this effort to control fertility that widespread and fundamental sexual ignorance was created. Women's more restrictive approach to sexuality is witness to the fact that for them sexual pleasure had been uncertain and at best transient, while the risk of pregnancy was constant and the impact of childbirth and a further child usually substantial.[50] Wright commented that where there is fear of pregnancy, 'anxiety holds the mind and sexual enjoyment becomes impossible'.[51] The manuals were the first explicit books about physical sexual activity written for an audience that included respectable, conventional women. Given the relative absence of other sources, it is probable that they played a major role in the dissemination of new and better birth control information.

The importance of birth control to the manual authors, or more generally doctors practising in sex, is reinforced by their involvement in the running of birth control clinics. Griffith and Wright were involved in setting up birth control clinics and also contributed to the campaign to legalize abortion.[52] Chesser worked as an illegal abortionist from the early 1940s until 1964 driven by 'profit and principles'.[53] There were other doctors involved in birth control and sex reform beside the manual authors discussed here. Dr Joan Malleson (born 1900) was doing sex research during the 1930s as well as running sex-counselling sessions at the Telford Road Birth Control Clinic. She belonged to the Worker's Birth Control Group in the 1920s, and helped found the Abortion Law Reform Association.[54] Dr Norman Haire who translated foreign sex manuals, contributed numerous prefaces, gave lectures on sexuality and helped found the British Society for Sex Psychology, also ran one of the first birth control clinics. The contribution of the manuals to the sexuality of heterosexual women in the first half of the twentieth century was undeniably positive. However, they did not have sufficient influence to create the new sexual identity which was coming into being. Effective birth control did.

Unlike later sexologists, such as William Masters and Virginia Johnson, the interwar medical manual authors saw feelings and motivations as central to sexuality. When Walker began life as a doctor, he explains, 'patients remained for me little more than collections of organs infused with life'.[55] Gradually he realised that 'feelings and emotions ... are states of being ... [that] affect every organ, every muscle, every tissue, every cell'.[56] However, his awareness of the mind did not result in a conventional interest in psychoanalysis. Freudian ideas permeated the manuals from the early 1930s, including Walker's own, but he added to this the ideas of the spiritual philosophers Gurdjieff and Ouspensky, and later became a founding member of the Ramakrishna and Vedantist Centre. Griffith had a more dogmatic and rigid character than Walker but he underwent a similar process, in his case precipitated by years in an unhappy marriage. He explains that 'when I was struck down with pneumonia I realised that the physical illness was but a pointer to a deeper inner disharmony which was disturbing the whole course of my life'.[57] This led him to an interest in Jung with whom he underwent several sessions of analysis in the 1940s. Griffith is an interesting, transitional figure. He rejected the prevailing British attitude that 'introspection is a bad thing', but only on the basis that introspection was actually a necessary part of emotional self-control.[58]

It is important to realize that many of the interwar manuals continued to sell into the 1960s and several of the authors were still alive. By the standards of the early and mid-twentieth century, these medical authors of sex manuals were tolerant and sympathetic towards homosexuals.[59] Their work contributed to lessening prejudice through the decades. In 1958 the Homosexual Law Reform Society (HLRS) was created and Kenneth Walker became the first chairman of the Executive Committee, and a founding trustee of the Albany Trust set up soon afterwards to promote the psychological health of the homosexual by collecting data and promoting research.[60] During the debates on the Homosexual Law Reform Bill, passed in 1967, heterosexual legislators had made it clear that homosexuals were expected to be discreet. Even supporters of the bill believed that the discretion required of homosexuals should be greater than that required of heterosexuals, and the law retained greater penalties for homosexual activity.[61] By the 1970s, a new younger generation of gay activists had emerged who saw the attitude of tolerance that the HLRS and its supporters promoted as almost worse than open rejection.[62] A Gay Liberation pamphlet rejecting psychiatry sums up their attitude to humane liberal tolerance:

> As [a] deviant group, we become increasingly resentful of this notion of tolerance for the unfortunate, which results in less savage punishment but also in a denial of our integrity and responsibility.[63]

They responded to tolerance as 'blatant, flaunting, determinedly iconoclastic, far-out, far-left sexual rebel[s]' insisting that the law reform had achieved little and demanding the right to express their sexuality openly.[64]

The discomfort with openly sexual behaviour by the generation of English people born in the late Victorian era extended beyond homosexuals (although the legal penalties did not) and, by the 1960s, even the most liberal among this generation appeared repressive. These manual authors' own personal lives appear to have been characterized by seriousness, discipline, dignity and emotional restraint – regardless of how unconventional they may have been. Helena Wright told her biographer that when marrying, 'she was wary of an emotional relationship and considered passion far too risky to contemplate'.[65] The researcher and homosexual Michael Schofield reviewed Helena Wright's *Sex and Society: a New Code of Sexual Behaviour* (1968). In spite of her extensive sexual experience and the discomfort that her support for 'extramarital sexual friendships' caused her publishers (and later her biographer), Schofield responded to the book by saying: 'It doesn't sound very enjoyable. Doesn't anyone have sex for fun?'[66] A few of the manual authors were responding to new ideas. When *Sex and the Married Woman* (1968) was published Eustace Chesser, the author, was aged sixty-six. In the manual he argued that 'the battle of the sexes is a long way from being over. Women have won a campaign but not a war.' A major theme of the book is that 'Modern woman ... rejects the whole concept that her husband has "conjugal rights" upheld by law, namely the right to enjoy her body whenever he wishes ... Her body is a piece of territory of which she has sole control.'[67] The text is a tribute to the man and fascinating to the historian but somewhat earnest and certainly not erotic as many of the earlier manuals, including Chesser's own, had been in their day. It was necessary to reject this generation as 'reformist' in order to fuel the ideological transition away from what had become, by the 1970s, a staid, legalistic, liberal tolerance to a more emotionally expressive, less contained sexual identity.[68] In the postwar period, heterosexual men also drew upon the sex manuals as authorities to support the imposition of normative standards of female sexuality upon women and the highly negative interpretation of the early manuals must be placed in the context of their deployment for this purpose in the 1950s.[69]

However, these interpretations are not relevant to an understanding of the early generation of medically trained British sex manual authors during the interwar period. There is more variation in attitudes among the other group of sex manual authors from the interwar period. The major non-medical British manual authors in the 1920s and 1930s included two clergymen, four professional writers and one scientist/writer (Stopes). The sex manual authors, Reverend Arthur Herbert Gray (born 1868) and Reverend Leslie Weatherhead (born 1893) shared an attitude of liberal *intolerance* to homosexuality:

> I find a certain amount of talk going on in some quarters which assumes that some people are of the homosexual type, and that it is natural and right for them to express themselves in this way. But this seems to be a complete and serious mistake.[70]

Weatherhead emphasised the difference between innate and acquired homosexuality. Those men and women whose homosexuality was acquired should simply desist, but he accepted that innate homosexuals endured much suffering and recommended they adopt the same procedures as those for stopping masturbation, which was regarded by him as a very serious problem.[71] In the interwar decades, even this degree of tolerance was unusual in a churchman and it was considerably more liberal than the attitudes of most British people.

Other authors focused on heterosexuality. Edward Charles (born 1898) took law at Oxford, lectured on English at foreign universities and wrote novels. His manual was not addressed to the 'working mothers', instead '[I am] consciously addressing myself to the most intelligent section of our society.'[72] He was interested in the 'healthy', 'normal' love of a man and woman, a relationship in which men were dominant and women were sexually responsive on demand (Charles believed D. H. Lawrence's Lady Chatterley to be too sexually demanding). He disliked the assertiveness required of women who used diaphragms and condoms diminished male pleasure, so, despite the evidence of ineffectiveness that existed by the mid-1930s, he recommended the use of spermicides alone, exposing women to a higher risk of pregnancy. George Ryley Scott, a writer who originally trained as an anthropologist, wrote over sixty books ranging from *The Art of Faking Exhibition Poultry, etc.* (1934) to *The History of Corporal Punishment: a Survey of Flagellation etc.* (1938). He felt less need than the doctors to be 'scientific' or to reassure his readers, especially the women. For example, in 1934, he explained that a woman's clitoris 'measures, in a state of normalcy, about one half inch in length but is capable of development if masturbation or excessive coitus is practiced'. '[B]estiality, like intercourse with a virgin, has a widespread reputation as a cure for venereal disease.' Not until the 1955 edition did he feel the need to mention that this cure did not work.[73] He provides a list of contraceptive methods to be used 'where the male partner wishes to prevent conception without consulting his partner', and an equivalent list for women.[74] Rennie Macandrew (born 1907) wrote simple short books that focused entirely on heterosexuals. He wanted women to be more sexually available (but was remarkably positive about masturbation, writing in 1938 that 'Gentle masturbation at intervals exercises and strengthens the genitals.').[75] The attitudes these men expressed towards women became increasingly dominant in the postwar period. Leonora Eyles (born 1889) was reduced to poverty before later success as a journalist. During this time she lived in a working-class area with her three children whom she had had before the age of twenty-four. She emphasizes many women's lack of sexual pleasure, and the huge importance of birth control in relation to this.[76] In her text, ignorance is not so central and problems of active female sexuality, such as extra-marital sexuality, are emerging.

The British doctors wrote more accurate, more compassionate and more open manuals than the non-medically trained male British authors in this period. Their own involvement in their material and their marginal status within the medical profession as doctors writing about sexuality gave them an incentive to understand and to accept others. Extracting the relevant anatomical information about the body from medical textbooks, creating descriptions of genitals and sexual acts, and choosing the information about contraception from a variety of sources required medical or, in the instance of Stopes, scientific training. Welding the whole together with advice and sentiment about marital relations in order to reject fear-inducing attitudes to sexuality required sensitivity. Even by the 1930s, when there were many models to copy, those authors who did not have medical training were more likely to produce books that were more prescriptive, less sensitive to the body and contained inaccurate advice.

The conceptual advances made by historians over previous decades have provided valuable analytical insights into normalization and the internalizing of new disciplinary techniques. Applying these insights in a nuanced and sensitive fashion does contribute to an interpretation of the manuals when they had developed into a normalizing discourse in the postwar period. However, this approach has now become a stifling orthodoxy. In the context of the current historiography on sexology, an essay such as this one, which accepts testimony by individuals as to their ideas and experience of sexual change and relates this positively to the developing discourse (as they did), feels like a revealing display of naivety. However, the challenge is to do precisely this; to integrate the insights that have been gained from discourse analysis over the past twenty years with the evidence of lived experience. That evidence suggests an account in stark contrast to the current historiography, one closer to much-derided notions of progress away from a state of ignorance and repression towards a society in which many people's experience of sexuality is happier and more satisfying, albeit certainly still not perfect. The historians of sexuality Richard Parker and John Gagnon acknowledge that 'No matter how much we now conceive these new doctrines as new forms of oppression and domination, the reformers and radicals of the time experienced them as liberating.'[77] Surely even 'we' might find it hard to see the following comments on sexuality as oppressive and dominating? These last words come from Kenneth Walker and Eustace Chesser, respectively:

> If the solution suggested by the writer is not one that commends itself to the reader, there will be no need for us to quarrel ... Truth has many facets and we see only a few of them.[78]

> *Happiness is the true test* ... Never mind what books – including this one – say you should do. If you are happy, and your partner is, too, *leave well enough alone.*[79]

## Notes

1. R. G. Parker and J. H. Gagnon (eds), 'Introduction', in *Conceiving Sexuality* (London: Routledge, 1995), p. 3; J. Weeks, *Sexuality and its Discontents* (London: Routledge & Kegan Paul, 1985), p. 78.
2. E.g. Weeks, *Sexuality*, pp. 94–5.
3. A. P. Pillay, 'Editor's Introduction', *Marriage Hygiene*, 1 (1947), 2. Dr Pillay did not give the names of the doctors visited. British doctors who contributed to the journal in its first year included Norman Haire, Joan Mallleson and Maurice Newfield.
4. All the sales figures given are worldwide and most have been taken from the dust jackets of the books. H. Wright's *The Sex Factor in Marriage*, 3rd edn (London: Williams & Norgate, 1930), 58 000 by 1940. G. C. Beale [pseud.], *Wise Wedlock: the Whole Truth* (London: Health Promotion, [1922]), 100 000 by 1939. Rev L. D. Weatherhead, *The Mastery of Sex: Through Psychology and Religion* (London: SCM, 1931), 58000 by 1942. Rev A. H. Gray, *Men, Women and God: a Discussion of Sex Questions from the Christian Point of View* (London: SCM, 1923), 93 000 by 1947. The most influential manual with the highest sales in the interwar period was *Married Love* (London: Putnam, 1918) by M. Stopes, which had sold 820 000 copies by 1937. Post-Second World War, the 1960 edition of R. Macandrew, *Life Long Love* (London:

Wales, 1938) claimed that his eight books had sold approaching 2 000 000 copies. According to her biography, Wright's *Sex Factor* had sold 1 000 000 by 1984. The publisher claimed T. H. van de Velde's *Ideal Marriage* (London: William Heinemann, 1928) had sold 700 000 by 1970. G. R. Scott's *The New Art of Love* (London: John Bale Sons & Danielsson, 1934) sold a relatively meagre 27 000 by 1954.

5. Letters to Stopes, see R. Hall, *Dear Dr Stopes: Sex in the 1920s* (London: Deutsch, 1978). Other authors: e.g. Beale, *Wise Wedlock*, p. 7; A. D. Costler and A. Willy, *The Encyclopaedia of Sexual Knowledge*, N. Haire, (ed.), 8th edn 1941 (London: Aldor, 1934), pp. 9–11; L. Eyles, *Commonsense About Sex*, 2nd edn, 1943 (London: Gollancz, 1933), p. 16; I. E. Hutton, *Memories of a Doctor in War and Peace* (London: Heinemann, 1960), p. 217; R. Porter and L. A. Hall, *The Facts of Life: the Creation of Sexual Knowledge in Britain, 1650–1950* (New Haven: Yale University Press, 1995), p. 283. In 1938, the Marriage Guidance Council was set up by Rev A. H. Grey, who was soon joined by Dr E. F. Griffith. By 1945, the organization was receiving letters 'at the rate of 4,000 a month', many of which were about sexual problems. J. H. Wallis, and H. S. Booker, *Marriage Counselling: a Description and Analysis of the Remedial Work of the National Marriage Guidance Council* (London: Routledge & Kegan Paul, 1958), p. 5.

6. H. Hopkins, *The New Look: a Social History of the Forties and Fifties in Britain* (London: Secker & Warburg, 1964), pp. 194–6.

7. E. Chesser, E. Cole, E. F. Griffith, I. E. Hutton, K. M. Walker and H. Wright. A. Havil was the pseudonym of Dr E. E. Phillip (b.1915). No information has been discovered about Dr G. C. Beale. The text suggests that the author was a well-educated man with medical training. L. Hall accepts Stopes' claim that he had plagiarized *Married Love* (1918) and describes him as an 'outright hack'. However, this claim seems to have been a product of Stopes' ego as the text of his manual is different in style to hers and he disagrees with her on several issues, so his manual has been included in this category. L. A. Hall, *Hidden Anxieties: Male Sexuality, 1900–1950* (London: Polity Press, 1991), p. 67. Porter and Hall, *Facts of Life*, p. 282.

8. E. Charles, L. Eyles, R. Macandrew (pseudonym of A. G. Eliot), G. R. Scott, M. Stopes, A. H. Gray and L. D. Weatherhead (assisted by Dr M. Greaves).

9. For an analysis including the foreign manuals see H. Cook, *The Long Sexual Revolution: English Women, Sex and Contraception 1800–1975* (forthcoming, Oxford, 2004).

10. M. Jackson, '"Facts of Life" or the Eroticization of Women's Oppression? Sexology and the Social Construction of Heterosexuality', in P. Caplan (ed.), *The Cultural Construction of Sexuality* (London: Routledge, 1987, repr. 1993). M. Jackson, *The Real Facts of Life: Feminism and the Politics of Sexuality, c. 1850–1940* (London: Taylor & Francis, 1994).

11. L. S. Lewis and D. Brissett, 'Sex as Work', *Social Problems* 158 (1967). M. Gordon and P. J. Shankweiler, 'Different Equals Less: Female Sexuality in Recent Marriage Manuals', *Journal of Marriage and the Family* (1971), 459–66. M. Gordon, 'From an Unfortunate Necessity to a Cult of Mutual Orgasm: Sex in American Marital Education Literature, 1830–1940', in J. M. Henslin and E. Sagarin (eds), *The Sociology of Sex* (New York: Schocken Books, 1978).

12. E. B. Freedman and J. D'Emilio, *Intimate Matters: a History of Sexuality in America* (New York: Harper & Row, 1988), p. 267.
13. Porter and Hall, *Facts of Life*, p. 221.
14. I. E. Hutton, *The Hygiene of Marriage*, ed. M. D. Smyth, 10th edn, 1964 (London: Heinemann Medical Books, 1923), p. 54.
15. K. M. Walker and E. B. Strauss, *Sexual Disorders in the Male*, 4th edn, 1954 (London: Cassell, 1939).
16. K. Wellings, J. Field, A. M. Johnson and J. Wadsworth, *Sexual Attitudes and Lifestyles* (London: Blackwell, 1994), T.6.8, p. 164, T.12, p. 175. D. Sanders, *The Woman's Book of Love and Sex* (London: Sphere, 1985), p. 81.
17. The most important sources are oral historians who have consistently found this the case when interviewing people born before the First World War. For example, see C. Chinn, *They Worked All Their Lives: Women of the Urban Poor in England, 1880–1939* (Manchester: Manchester University Press, 1988); S. Humphries, *A Secret World of Sex* (London: Sidgwick and Jackson, 1988); D. Gittins, *Fair Sex, Family Size and Structure, 1900–39* (London: Tavistock, 1982); E. Roberts, *A Women's Place: an Oral History of Working-class Women, 1890–1940* (Oxford: Blackwell, 1984); P. Thompson, *The Edwardians: the Remaking of British Society* (London: Weidenfeld and Nicolson, 1975). Other evidence includes R. Roberts, *The Classic Slum: Salford Life in the First Quarter of the Century* (Manchester: Manchester University Press, 1971); L. England, 'A British Sex Survey', *International Journal of Sexology*, 3, 3 (1950); M. Cole, *The Life of G. D. H. Cole* (London: Macmillan, 1971), pp. 89–94; J. Calder, *The Nine Lives of Naomi Mitchison* (London: Virago, 1997), pp. 41–3. Concern to prevent sexual material being made available to women, and young or working-class people, was expressed, for example, during the interwar trials relating to the work of M. Stopes, Radclyffe Hall, E. Charles and E. Chesser.
18. L. Jordanova, 'The Social Construction of Medical Knowledge', *Social History of Medicine*, 8 (1995), 369.
19. M. Foucault, *The History of Sexuality* (*La Volonté de Savoir*) (London: Peregrine, 1978), p. 84.
20. Foucault, *History*, pp. 17–8.
21. *Childhood Sexual Learning: the Unwritten Curriculum*, ed. by E. J. Roberts (Cambridge MA: Ballinger, 1980).
22. See Cook, *Long Sexual Revolution*, chs 6 and 7.
23. Wright, *Sex Factor*, 3rd edn, p. 92. The belief that shame at masturbation was natural was widely accepted.
24. E. F. Griffith, *The Pioneer Spirit* (Upton Grey: Green Leaves Press, 1981), p. 23.
25. Griffith, *Spirit*, p. 28.
26. K. M. Walker, *Sexual Behaviour: Creative and Destructive* (London: Kimber, 1966), p. 185.
27. E. F. Griffith, *Modern Marriage*, 19th edn, 1947 (London: Gollancz, 1935), p. 131.
28. Hutton, *Memories*, p. 214. For a brief discussion of Hutton's manual, see Hall, *Hidden Anxieties*.
29. See e.g. D. Russell. *The Tamarisk Tree: My Quest for Liberty and Love* (London: Virago, 1975).

30. B. Evans, *Freedom to Choose: the Life and Work of Dr Helena Wright, Pioneer of Contraception* (London: Bodley Head, 1982), p. 152.
31. Walker, *Sexual Behaviour*, p. 117.
32. Walker, *Sexual Behaviour*, p. 185.
33. Griffith, *Pioneer Spirit*, pp. 23, 28.
34. P. Ferris, *Sex and the British: a Twentieth Century History* (London: Mandarin, 1994), pp. 266–8. E. Chesser, *Love Without Fear (A Plain Guide to Sex Technique for Every Married Adult)* (London: Rich & Cowan, 1941).
35. Jackson, *Real Facts*, p. 163. Jackson, 'Facts of Life', pp. 60–1. See also S. Jeffreys, *The Spinster and Her Enemies: Feminism and Sexuality, 1880–1930* (London: Pandora, 1985).
36. Jackson, *Real Facts*, p. 185, see also p. 75.
37. Jackson, *Real Facts*, pp. 163–4. Jackson, 'Facts of Life', pp. 60–1.
38. Stopes, *Married Love*, 18th edn, p. 65. Jackson, *Real Facts*, ch. 6.
39. Self-proclaimed sexologist E. Charles wrote that when he saw copies of *Married Love* for sale he thought of: '[T]he big green mantis sitting up on her hind legs and eating husband after husband as soon as he has satisfied her biological needs'; E. and M. Charles, *Indian Patchwork* (London: Heinemann, 1933), pp. 283–5. See also E. Charles, *The Sexual Impulse* (London: Boriswood, 1935), pp. 9–11. D. H. Lawrence wrote an abusive essay about Stopes. Thanks to her, he wrote, sex had been 'mentalised', it was all 'cerebral reaction', and 'not a bit of phallic insouciance and spontaneity'. H. T. Moore (ed.), *Sex, Literature and Censorship* (London: Heinemann, 1955), p. 76. K. Briant, Stopes' former lover, wrote that 'she was a woman writing with a male personality and a male approach to the sex act'. Quoted in L. Taylor, 'The Unfinished Sexual Revolution: the First Marie Stopes Memorial Lecture', *Journal of Biosocial Science*, 3 (1972), 474. See also R. A. Soloway, '"The Perfect Contraceptive": Eugenics and Birth Control Research in Britain and America in the Interwar Years', *Journal of Contemporary History*, 30 (1996), 643. Most recently, P. Ferris has written, 'she was, perhaps, not quite sane', *Sex and the British*, p. 109.
40. N. Mitchison described the contribution reading *Married Love* made to her marriage in these terms. Calder, *Mitchison*, pp. 42–3, 48, 92–3. Eyles, *Commonsense*, p. 46.
41. Jackson, *Real Facts*, p. 151. Quoting Frances Prewett, in *The Woman's Leader*, 12 May 1921.
42. Ibid., p. 135.
43. P. Shapiro, 'The Unplanned Children', *New Society* (November 1962). This argument is also put by E. Wilson, *Only Halfway to Paradise: Women in postwar Britain, 1945–1968* (London: Tavistock, 1980), p. 99. (The Shapiro reference is missing from the bibliography.) K. M. Walker and O. Whitney, *The Family and Marriage in a Changing World* (London; Victor Gollancz, 1965), pp. 132–4. See also L. Gordon, 'Why Nineteenth-century Feminists did not Support "Birth Control" and Twentieth-century Feminists do: Feminism, Reproduction and the Family', in *Rethinking the Family: Some Feminist Questions, ed.* B. Thorne and M. Yalom (London: Northeastern University Press, 1992).
44. M. Stopes, *Wise Parenthood* (London: Fifield, 1918).

45. The exception was Wright who recommended Fielding. Wright, *Sex Factor*, 3rd edn, p. 96. M. Fielding, *Parenthood: Design or accident? A manual of Birth Control* (London: The Labour Publishing Company, 1928).

46. Griffith, *Modern Marriage*, 19th edn, p. ix. This was an almost completely rewritten version of his 1935 book, the greater part of which had been given over to birth control.

47. E.g. Th. H. van de Velde, *Fertility and Sterility in Marriage* (London: Heinemann, 1931), pp. 300–1; M. Stopes, *Contraception, Theory, History and Practice* (London: Bale & Danielsson, 1923), pp. 138–9, 61–3.

48. K. M. Walker, *The Physiology of Sex* (London: Penguin, 1940), p. 69.

49. S. Szreter, *Fertility, Class and Gender in Britain, 1860–1940* (Cambridge: Cambridge University Press, 1996). Thompson, *Edwardians*, p. 78. A similar argument has been made regarding the USA. P. A. David, and W. C. Sanderson, 'Rudimentary Contraceptive Methods and the American Transition to Marital Fertility Control, 1855–1915', in S. L. Engerman and R. E. Gallman (eds), *Long Term Factors in America's Economic Growth* (Chicago: University of Chicago Press, 1986).

50. See Cook, *Long Sexual Revolution*, ch. 1.

51. H. Wright, *Sex Fulfilment in Married Women* (London: Williams & Norgate, 1947), p. 73.

52. Griffith, *Pioneer Spirit*, pp. 46–7, 57, 67. Evans, *Freedom*, pp. 247–8.

53. On Chesser, see Ferris, *Sex and the British*, pp. 266–8.

54. S. Brooke, '"A New World for Women"? Abortion Law Reform in Britain During the 1930s', *American Historical Review*, 106 (2001). L. Hoggart, 'The Campaign for Birth Control in Britain in the 1920s', in A. Digby and J. Stewart (eds), *Gender, Health and Welfare* (London: Routledge, 1996). Evans, *Freedom*, p.148.

55. K. M. Walker, *A Doctor Digresses* (London: Cape, 1950), p. 17.

56. K. M. Walker and P. Fletcher, *Sex and Society* (London: Penguin, 1955), p. 136. Also see, Walker, *Sexual Behaviour*, p. 171.

57. Griffith, *Pioneer Spirit*, p. 101.

58. Griffith, *Pioneer Spirit*, pp. 102, 104. He may have first encountered Jung when the latter gave a seminar in England in 1923. (p. 23). By April of 1948, when he went to Zurich for six weeks and had a few private sessions with Jung, he had already rewritten *Modern Marriage* (1935) to include a chapter on emotion.

59. K. Walker (ed.), *Preparation for Marriage: a Handbook Prepared by a Special Committee on Behalf of the British Social Hygiene Council* (London: Cape, 1932), pp. 152–3. H. Wright, *Sex and Society* (London: Allen & Unwin, 1968), pp. 80–1, 84.

60. A. Grey, *Quest for Justice: Towards Homosexual Emancipation* (London: Sinclair- Stevenson, 1992), pp. 27–8, 30. E. Chesser's *Live and Let Live: the Moral of the Wolfenden Report* (London: Heinemann, 1958) was distributed by the HLRA. H. Montgomery Hyde, *The Other Love* (London: William Heinemann, 1970), pp. 6, 169.

61. Lord Arran, who piloted the homosexual law reform bill through the House of Lords said: 'I ask those who have, as it were, been in bondage and for whom the prison doors are now open to show their thanks by comporting themselves quietly and with dignity'; cited in J. Weeks, *Sex, Politics and*

*Society*, 2nd edn (London: Longman, 1981), p. 274. Lady Gaitskell said in the House of Lords: 'When I see two young people embracing on the moving staircase on the tube I am not outraged myself, but I should be outraged if I saw two homosexuals doing just that'; quoted in Grey, *Quest For Justice*, pp. 108–9.

62. Grey, *Quest*, p. 179.

63. *Psychiatry and the Homosexual: a Brief Analysis of Oppression*, Gay Liberation pamphlet no. 1, cited in Grey, *Quest*, p. 181. See also J. Weeks, *Coming Out: Homosexual Politics in Britain, from the Nineteenth Century to the Present* (London: Quartet, 1977). B. Cant and S. Hemmings (eds), *Radical Records: Thirty Years of Lesbian and Gay History, 1957–1987* (London: Routledge, 1988).

64. Grey, *Quest*, p. 183.

65. Evans, *Freedom*, p. 79.

66. M. Schofield, 'Review: *Sex and Society* by Helena Wright', *Family Planning* (1968), Evans, *Freedom*, pp. 170–1.

67. E. Chesser, *Sex and the Married Woman* (London: Allen, 1968), pp. 13, 118.

68. For the academic construction of the 1960s legal and social changes as 'reformist' and supportive of the capitalist system, see *Permissiveness and Control: the Fate of the Sixties Legislation, 1977 Sheffield* (London: Macmillan, 1980). See also J. Weeks, 'Sexual Values Revisited', in L. Segal (ed.), *New Sexual Agendas* (London: Macmillan, 1997), pp. 43–59. A. Grey describes how he himself was rejected as reformist in the early days of the GLF – but in his gratitude to those heterosexuals who accepted him and supported law reform he expresses precisely the deferential attitude that the younger generation rejected. Grey, *Quest*.

69. E.g. see 'I myself have clear memories of them being read and discussed by students when I was an undergraduate in the late 1950s'; Jackson, *Real Facts*, p. 156.

70. Gray, *Men, Women*, p. 81. Weatherhead, *Mastery of Sex*, pp. 151–8. Stopes expressed similar attitudes. M. Stopes, *Enduring Passion*, 4th edn, pp. 40–1.

71. Weatherhead, *Mastery of Sex*, pp. 134–9, 158. Griffith reprinted Weatherhead's advice on stopping masturbation in his first manual but commented only on homosexuality in adolescence. He felt this was acceptable, although physical expression of the emotions was to be resisted by boys, and girls should avoid friends who were too demanding.

72. E. Charles, *Sexual Impulse*, p. 155.

73. Scott, *New Art*, pp. 27, 110, 4th edition, p. 85.

74. G. R. Scott, *Modern Birth Control Methods*, 2nd edn, 1947 (London: Torchstream Books, 1933), pp. 30, 31, 33.

75. Macandrew, *Life Long*, pp. 28, 84.

76. Eyles claimed that in 1941, 'the only year of which I kept any statistics', over 7000 women wrote to her telling her they 'either detested and refused to have intercourse or put up with it with boredom, or longed to find in it something to enjoy'; *Commonsense*, p. 16.

77. Parker and Gagnon, 'Introduction', p. 5.

78. Walker, *Physiology*, p. vi.

79. Chesser, *Love Without Fear*, p. 94.

# 12

# Hailing a Miracle Drug: the Interferon

*Toine Pieters*

In the 1970s cancer therapy became the focus of the increasingly politically powerful, critical health movement in America, highlighting the failures and severe side-effects of conventional cancer treatments: surgery, radiation and chemotherapy. Faced with unprecedented public criticism of the low success rate of the costly 'battle against cancer', the American cancer establishment was seriously under attack. It needed a scientific promise that suited the growing public demand for effective and less toxic cancer remedies. The politically well-informed supporters of a new type of cancer agent, named 'interferon', capitalized on this need and the popular desire for more natural and organic remedies. In publicly emphasizing the presumed non-toxic and natural qualities of interferon as both an unorthodox organic and science-based promise in the fight against cancer, they succeeded in getting interferon absorbed in the accelerating politics and economics of the American cancer scene. With the press blossoming into the most important agenda-setting forum, a boom in expectations was fuelled world wide, resulting in interferon acquiring 'miracle drug' status in the late 1970s.

However different its therapeutic properties, interferon shares this special but transitory 'wonder drug' label with penicillin, cortisone, streptomycin and, lately, viagra. They are among the elite group of twentieth-century therapeutic drugs that have succeeded in achieving this divine quality. As such they have taken on special symbolic significance, as icons of a religious belief in the healing power of modern medicine. In recent years a growing number of medical historians have elaborated on the complex socio-cultural phenomenon of the 'miracle drug'. David Cantor was the first to try opening up the 'black box' and recover the frame of action and meaning related to this powerful phenomenon. In his study of cortisone, he describes how the cycle of promise and disappointment is closely connected with the commitments and interests of doctors, pharmaceutical companies and the state.[1] Cantor shows how all these different parties were involved in a so-called 'politics of drama'. Basically this means that cortisone's status as a wonder drug resulted in part from common interests in publicly exploiting its dramatic qualities as a potentially golden asset to the doctor's bag. Allan Yoshioka's account of streptomycin is different in scope. Without neglecting the politics of drama, he emphasizes the cultural significance of the phenomenon as a transient product of a religious faith in modern medicine which promises to fulfil the quest for immortality.[2] Robert Bud's analysis of the story of penicillin, in turn, uses the mythology of a wonder drug to point out the important symbolic role that is played by such imaginative products of modern science and technology, serving not only to carry messages of hope and promise, but also as a lubricant for future scientific endeavours.[3]

In this profile of a miracle drug, the focus is on reconstructing and analysing the cultural process through which the status of wonder drug was achieved in the case of interferon. Special attention will be paid to the channels of communication and mediation. Major actors who influenced the construction and circulation of interferon both as individuals and as a collective ranged from doctors, laboratory researchers, patients, journalists and regulators, to drug company executives, and they did so within the context of mass culture in the 1970s. In the conclusion I will also point out the similarities and differences between interferon and earlier carriers of the wonder drug emblem – penicillin, streptomycin and cortisone – in terms of how their breakthrough imagery entered circulation in mass culture and how in the process this imagery successively intensified and weakened.

## The emergence of interferon as an anti-cancer agent

In the spring of 1958, less than a year after the British virologist Alick Isaacs and the Swiss medical researcher Jean Lindenmann announced their collective discovery of a biological factor with distinctive antiviral properties named 'interferon', interest in interferon as a therapeutic agent started to grow in Britain. As a champion of persuasive public demonstrations and a strong competitor for public credibility Isaacs played a pivotal role in bringing interferon, however unsubstantiated its therapeutic effects, to the public stage. Isaacs promoted interferon as an innocuous biological substance that might offer a potential new approach to the medical treatment of virus infections. This message was not lost on the British Press. *The Daily Telegraph*, among other newspapers, depicted interferon in terms of the discovery of a new 'antiviral penicillin'.[4] According to the *British Medical Journal* interferon broke new ground in medicine as a revolutionary therapeutic tool against viral disease appropriate to the satellite age.[5] Isaacs was even asked to give a demonstration of interferon's workings on the BBC programme *Eye on Research*.[6]

An important incentive for Isaacs' success at connecting his private science with public issues of disease and cure was the strong public desire for and belief in emulating the success of penicillin, streptomycin and cortisone. The public portrayal of their spectacular clinical performances had raised hopes regarding scientific medicine's ability to come up with yet another 'wonder' at the disposal of the doctor. Of course this belief, far from being unique to the British, was shared by Americans and others alike. But what was special in the British context was the conjunction between a firm belief in breakthroughs as a dominant feature of scientific medicine and the postwar penicillin trauma. The latter was due to an inability to control the patents; British doctors had to pay royalties to American firms for using penicillin, a British discovery. This resulted in interferon being touted publicly in Britain. Without additional demonstrations of dramatic therapeutic results and with accumulating purification and production problems, however, the British euphoria for interferon as a potential cure for viral disease lapsed. By the mid-1960s interferon had earned itself the telling nickname 'misinterpreton' and was relegated to the category of obscure laboratory substances. But living in obscurity would turn out to be a temporary sidestep in a shifting biography.[7]

In the USA in early 1970, Mary Lasker, the philanthropist and notorious lobbyist of American medical research, had assembled a panel of consultants. This group, largely consisting of politically influential associates and friends also known as the 'Laskerites or Mary's little lambs', was to advise the American Senate on legislation involving a 'moonshot' approach for cancer that held out the promise of major progress in the officially declared 'war on cancer'.[8] One of the scientific panel members was Mathilde Krim, a Swiss-born geneticist and virologist at New York's Memorial Sloan Kettering Cancer Centre, whose husband Arthur Krim, a media tycoon, was influential in the Democratic Party. Krim helped draft working papers on the progress of cancer research, which became part of the technical portion of the panel's report entitled *National Program for the Conquest of Cancer.*

Krim did her share of the work in searching the literature for promising areas of cancer research that deserved more attention. In doing so she came across the interferon literature and learned that studies had been published claiming that interferon, by means of its antiviral activity, had an inhibitory effect on tumour viruses and virus-induced tumours in mice and rats. In accordance with the mainstream in virology Krim firmly believed that tumour viruses played a major role in the etiology of malignancies not only in animals but also in humans. Unlike most of her scientific colleagues, however, she disregarded the 'misinterpreton' rumours and qualified interferon as a potentially interesting anti-tumour agent for use in humans. Krim managed to include interferon research in the panel's report as a promising area that needed intensive further study.[9]

After presentation of the report to the Senate, Krim remained closely involved with the pre-legislative agenda-setting activities of Mrs Lasker up to the enactment by President Richard Nixon of the National Cancer Act in December 1971. This event marked the start of a much-expanded National Cancer Program and was the outcome of more than a year of political struggle and compromise in which Mathilde Krim became familiar with the ins and outs of American cancer politics. According to Krim it was Lasker who taught her the political groundwork of science lobbying and the persistence needed to succeed as a self-appointed lobbyist.[10] In the case of interferon not a little endurance would be required. Despite her lobbying skills and her good connections it took Krim years to find an audience for interferon as a serious lead towards cancer therapy.

Sandra Panem has already pointed to some of the factors that worked against Krim: apart from being a woman in a 'man's world', Panem noted that Krim was regarded as an outsider in the field of interferon research, neither knowledgeable enough nor sufficiently critical about interferon. Among interferon researchers and the National Institutes of Health bureaucracy there was anxiety that the outsider Krim might gain too much control over the future course of interferon research.[11] Another prominent factor was resistance to the idea that the action of interferon might be pleiotropic, with a therapeutic effect both against viruses and against cancer. This did not fit the notion of specific etiology of disease and of specific therapy on which most therapeutic drug research programmes were based. Moreover, there were still serious doubts about the scientific status of this poorly defined and impure biological substance for which an econ-omically feasible purification and production technology was not yet available.

Krim nevertheless succeeded in managing the opposition effectively. To her, interferon represented as much politics as science. As a political animal with a feel for the changing spirit of the age Krim tried to capi-talise on the growing public dissatisfaction with the effectiveness and side-effects of conventional cancer treatments: surgery, radiation and chemotherapy. The politically influential critical health movement openly accused the patrons of American cancer research of wasting tax-payers' money and neglecting the search for the natural and organic.[12] In placing a great deal of emphasis on the natural and presumed non-toxic qualities of interferon as an unorthodox product of scientific medicine, Krim managed to address both critics and supporters of the National Cancer Program and build support for her 'interferon crusade'.

Once interferon became absorbed into the accelerating politics and economics of the American cancer scene, it was co-opted by a growing number of social groups. Apart from shared expectations about interferon as a new form of cancer treatment that was non-toxic, each group had specific motives for jumping on the interferon bandwagon: from the group of interferon researchers and the advo-cates of immunotherapy, who came to realize that Krim's interferon 'campaign' might be helpful in giving a new impetus to their research fields; to the genetic engineers, who were looking for suit-able demonstration projects to prove the commercial worth of the rDNA technology. Moreover, the patrons of cancer research were seriously under attack and needed a boost in the fight against cancer to satisfy public demands. As a product of scientific medicine, and with its claimed naturalistic basis, interferon seemed to offer the opportunity to silence the growing opposition. According to Krim, 'The American Cancer Society and some of the others felt they would really like to have something dramatic happen ... they believed they were making progress but not enough to satisfy the public.'[13] So in July 1978 Frank Rauscher, who had resigned as direc-tor of the National Cancer Institute (NCI) to become the American Cancer Society's vice-president for research, jumped at a request for funding of interferon research. In his previous post at the NCI the very same Rauscher had opposed funding interferon research broadly speaking on the basis of the same clinical test reports, but that was in 1975 when Krim's 'interferon crusade' still had to gain momentum.

## Interferon, scientists and the media

On 30 August 1978, the American Cancer Society (ACS) publicly announced it would spend $2 million on clinical studies with interferon to investigate its value in treating advanced cancer. This was the largest grant the ACS had ever committed to a single project. The fact was obviously not lost on the media as the press release triggered a wave of interferon-related publicity with headlines such as 'Interferon: the cancer drug we have ignored', 'Test planned on substance used in cancer treatment', 'New cancer weapon?', or 'Natural body substance: $2 million test on cancer retardant'.[14] Together with additional interest shown by radio (e.g. by the National Public Radio news programme *All Things Considered*) and television networks (e.g. ABC, CBS and NBC news) this brought interferon, as we shall see, out of the relative seclusion of the laboratory into the limelight of public attention.[15]

Krim played a pivotal role in shaping the public image of interferon. She was well aware of the ever-growing importance of the mass media in mobilizing massive support and was very skilful at 'doctoring' the media in ways that would fulfil the public desire for cancer cures. She and her fellow 'interferon crusaders' helped to create a media image of interferon as a somewhat mysterious, clinically unharnessed, non-toxic natural body substance, that offered a unique, though long-ignored, line of biomedical research. Their double framing of interferon as a natural solution to a dread disease and as a product of 'cutting-edge' biomedical research was reinforced by the illustrations employed in the media. A case in point was a photograph in *Newsweek* with the caption, 'Searching for the natural key to interferon': it showed Krim in a laboratory coat staring hopefully at a sophisticated laboratory set-up composed of a tangle of wires, tubes, retorts and graduated cylinders filled with fluids and suggested only experts can understand such complicated machinery.[16] This kind of public imagery served to link state-of-the-art laboratory research with the achievement of future bedside cures.

The picture conveyed most often and most vividly in the mass media was that if only enough of this extremely scarce and expensive naturally occurring protein could be made available by scientists, some kind of miracle cure was at hand for everything from cancer to the common cold. Interferon had all the ingredients for producing

dramatic leads, such as: 'Interferon: no miracles without more molecules', or 'If IF works it could ... be a vital weapon in the battle against
cancer, protect against viral and bacterial disease, provide a cure for
shingles, rabies, chicken pox, eye infections and prevent the common
cold', 'It attacks viruses, is tested in cancer labs; but it is scarce and
costly' and 'Will interferon kill cancer? Finnish Dr Kari Cantell is
helping the world find out'.[17]

Despite the sobering facts about interferon's questionable therapeutic exploitability that underlay these bold headlines, the frequent association of interferon with terms like 'cancer', 'weapon', 'natural body
substance', 'miracle' or 'scarce and costly' fuelled public expectations.
The fact that this potential cancer therapy was costly and extremely
hard to come by, together with the knowledge that no other biological
substance could match interferon's extraordinary biological activity, all
contributed to the popular belief that interferon must be highly effective. Most of these descriptions in the media were directly or indirectly
inspired by enthusiastic accounts of research events by biomedical
researchers. According to *Washington Post* columnist Nicholas von
Hoffman the uncertainty about interferon's effectiveness as an anti-
cancer cure did not discourage 'any number of persons in white
smocks from making their debuts before the cameras to conjure up
rose-hued dreams of therapeutic miracles'. Of course they did their best
to pepper their public declarations with 'if', 'could' and 'might'. But
the mere fact that interferon was taken seriously by a scientific community that was highly critical of other unorthodox cancer remedies,
like vitamin C and laetrile, was already enough to let the message of
promise and hope stick with audiences.

This public message of hope was reinforced by accompanying,
appetite-whetting stories in the media. Reporters included rich and
famous cancer patients such as the movie star John Wayne, Ted
Kennedy's son and the exiled Shah of Iran to magnify their interferon stories.[18] In doing so they stimulated further public speculation
about the therapeutic potential of this natural substance that was
extremely hard to come by. This and references to interferon in the
media, such as 'interfering with cancer' and 'the body's own wonder
drug', paved the way for the popular belief that interferon was highly
effective, but also like a magic potion that came close to being 'God's
own elixir'.[19]

Claims by genetic engineering firms and leading molecular biologists
about the imminent possibility of making available, in large quantities
and at modest cost, this potential breakthrough in the fight against

cancer intensified public interest.[20] According to *The Observer* the race was 'on for miracle drug'.[21] The fierce scientific competition involved in the efforts to clone the interferon gene was compelling in its own right, but not because it was one of the first medically significant human genes to be cloned. Nor because it proved that genetic engineering had passed, as the *Wall Street Journal* put it, 'from science-fiction fantasy to fact'.[22] Rather, it was the metaphor of a race, in combination with both the promise of wonderful new production technology and of a wonder drug in the making with a value beyond price, that made interferon so fascinating to the general public.[23]

Public euphoria reached a peak after the genetic engineering firm Biogen and the pharmaceutical company Schering-Plough Corporation announced at a joint press conference at the Park Plaza Hotel in Boston on 19 January 1980 – skilfully orchestrated to arouse media attention – that by genetic engineering the Swiss molecular biologist Charles Weissmann and his team were the first to succeed in getting bacteria to produce human interferon in biologically active form. While admitting that there were still a lot of questions to be answered, such as the extent to which the rDNA-made interferon was different from the natural interferon, Nobel Prize winner Walter Gilbert, a Harvard professor and chairman of the board of Biogen, predicted that within one or two years mass production of interferon for use in clinical trials would be feasible.[24]

During the days and months that followed the story appeared on the front pages of most American newspapers and magazines, and other media in Europe and also in Asia, Africa and South-American countries ran major pieces on it.[25] The major networks too led their prime time newscasts with the latest promising news on interferon. Readers and viewers were bombarded with headlines such as 'Medical breakthrough reported: "glamour stock" could help cancer patients', 'Cancer treatment available soon', or 'The big IF in cancer: will the natural drug interferon fulfill its early promise?', and 'At only $100 million a gram, this "miracle" has a future'.[26] These kinds of leads were supported by familiar yet powerful images of scientists and doctors in white coats busy working with tubes and syringes in their sophisticated, high-technology laboratories and clinics. Interferon hype spread worldwide. The events were described by Nicholas Wade in *Science* in terms of a 'cloning gold rush' that turned molecular biology into big business – with biotechnology stocks rising to record levels on the international stock markets. Popular magazines talked about genetic engineering as the solution to the problem of producing a 'priceless miracle drug'.[27] All sections of society played a part in promoting interferon to the elite league of wonder cures. As Sandra Panem aptly portrayed the situation:

> ... scientists who genuinely believed that they were on the right track and that money solicited at the expense of candour would be wisely used; investors and the public who wanted interferon to be a wonder drug and did not choose to ask whether the claims might be overstated; and those representatives of the media who reported anecdotes with unbridled enthusiasm.[28]

The dramatic portrayal of interferon seemed to work to the advantage of all, with enormous amounts of energy and money pouring into efforts related to interferon in times of otherwise sharp financial cutbacks. Developments such as federal and industrial research programmes, the involvement of major immunology and molecular biology laboratories, large-scale multi-centre trials, and the start of both an International Society for Interferon Research and the *Journal of Interferon Research* all occurred rapidly.

However, there was another side to the hype. Cancer centres in America and Europe were inundated with requests for interferon. Countless desperate patients and their families were begging hospitals, doctors, research centres and drug companies to provide them with the new wonder drug. This kind of interferon mania only added to media interest: 'Wonder drug hope of Miss Anneli: amazing case book of the wonder drug doctor', 'Dad's wonder drug plea' and 'Drug brings hope for tumor boy Daniel'.[29] The interferon story contained the drama of life and death, the horror of dying patients, the awesome picture of decades of obscure and difficult research, and the dedication and persistence of a handful of scientists to produce minute amounts of a potential life-saver. Personification of scientific medicine and disease – projecting the inherent benevolence of medical science on to individual scientists, doctors and patients – was a powerful rhetorical tool that journalists routinely employed. Imagery got inextricably bound up with content and this helped to intensify the interferon furore.

This can be exemplified by looking at parts of the British ITV programme *TV Eye* entitled 'Cancer: the new weapon' that was broadcast on 19 June 1980.[30] The documentary began with a close-up of an ampoule of interferon which the voice-over 'modestly' introduced as follows:

This phial of medicine contains one of the rarest and most expensive substances in the world. It's also one of the most exciting. It's already been used successfully to treat some cancer patients ... which could offer hope to millions of cancer patients throughout the world. We report on its successes, its failures and the controversy surrounding its use.

The programme proceeded with a portrait of four apparently successfully treated cancer patients. First came the same 16-year-old Swedish girl, Anneli, who figured prominently in the British tabloid *The Sunday Mirror* shortly before, as the happy, dance-loving and horse-riding 'interferon miracle'. While the camera zoomed in on an exultant looking Anneli the voice-over informed the viewer that this teenager, with a highly malignant form of cancer, owed her life to interferon. 'Six years ago her chances of surviving a cancer operation were one in five. Today, she is free of cancer.' Although it was emphasized that the doctor involved, the leading world expert on interferon Hans Strander, remained cautious, his photograph, showing a young intently looking but self-effacing medical scientist, rather dispelled this doubt. And interferon's startling abilities to heal ensured that the question, 'Does interferon work against cancer?', became more and more rhetorical in nature. Add to this the eager reproduction of the powerful penicillin analogy and the viewer seemed to have no choice other than to believe that interferon must work – if not yet then certainly in the near future – in spite of all the doubts still circulating among medical professionals.

If it lives up to its promise and as its name implies, that's still a very big IF, its discovery will rank with Salk's development of the polio vaccine and the discovery of penicillin by Alexander Fleming [accompanied by familiar library pictures of these medical heroes]. In fact there are many similarities between the story of penicillin and that of interferon. Both were discovered in Britain but developed elsewhere. Fleming's discovery was also expensive at first, now it's cheaper than the bottle it's put in. Fleming's use of penicillin was surrounded by controversy. So too is interferon. (*TVEye*, 'Cancer: the new weapon', 19 June 1980)

The reiterative nature of the public comparison between interferon's journey from laboratory to clinic, with the long, obstacle-filled road that penicillin had travelled from Fleming's laboratory to the pharmacist's shelf, lent not only additional credibility to interferon's effectiveness but also reinforced its status as a miracle drug.[31] In the British context these inflationary forces were stronger than anywhere else due to a specific cultural condition, the 'penicillin trauma', which also played an important part in the public reception of interferon as 'the antiviral penicillin' in the early 1960s. Inadvertently or not, the programme capitalized on the slumbering public sentiment that once again Britain lost out by failing to exploit its own discoveries, with the following statement:

> So, throughout Europe and America the pace of interferon research, development and testing is moving ahead very quickly. But what of Britain? Where after all interferon was originally discovered. Well sadly, as so often happens with British inventions, Britain is now lagging behind the rest of the world ... The Imperial Cancer Research Fund is still negotiating for supplies and cannot start clinical trials until autumn. Last year they collected more than twelve million pounds in donations from the British public. At this stage they refused to discuss their plans. (*TVEye*, 'Cancer: the new weapon', 19 June 1980)

Add to this the postscript 'please do not ask your doctor about interferon because supplies just do not exist' and the viewer is left with the impression that the only barriers to interferon's application were financial support from a sluggish government and opposition from a conservative medical establishment.

In the light of media messages and imagery prone to speculation of this kind, it is unsurprising to find that there was massive demand from patients suffering from cancer or a severe viral disease wanting access to this 'miracle drug' despite its being in short supply and as yet untested. Much distress was involved. In their face-to-face contact with patients, doctors found themselves besieged by demands for a drug that they could not supply, and frustrated by hopes of a cure they could not deliver. At centres where interferon was being tested everyone wanted to be in the trial and doctors had a hard time explaining that nobody could or would be favoured in their selection of trial candidates. In order to determine whether interferon had any 'real' efficacy medical researchers had to work in accordance with stringent testing protocols which implied that only patients meeting the highly specific trial requirements would be allowed to participate. And given the test subjects' own enthusiasm for interferon in combination with their special social status as lucky dogs this was already asking for the moon. The doctors involved in testing interferon feared that the high hopes of 'the chosen' might work against what was defined as 'objective benefit' and as such might undermine the scientific validity of interferon.[32]

Confronted with the disturbing side-effects of the hype, the biomedical establishment increasingly regarded the media frenzy as problematic. Medical research organizations like the British Medical Research Council and the Imperial Cancer Research Fund (ICRF) and health administrators, saw the public demand for interferon as a challenge to their authority of drug evaluation, registration and supply procedures.[33] Media stories about distressed families of terminally ill cancer patients willing to sell their homes and businesses to pay for their loved ones to be treated with interferon, led to worries that manufacturers would yield to public pressure to provide interferon outside the formal testing channels and, worse still, that a black market for interferon might develop.[34] Medical authorities realized that it was hard to explain to a patient suffering from cancer that there might be a more effective treatment in the pipeline but that it would not be generally available for some years, until licence procedures had been completed, by which time he or she might be dead.[35]

Distribution of the unlicensed drug interferon outside the approved medical trials was not only believed to undermine the ongoing drug evaluation process but also the whole state-regulated drug testing practice – established in the 1960s to maintain certain quality and safety standards.[36] With the prospect of improving supplies it seemed even more difficult to resist the public's disinclination to wait and to ensure that all available material would be channelled into the official trajectory of controlled clinical studies, required for the proper evaluation and licensing of new drugs.[37] Confronted with the desperate efforts of families and friends of cancer sufferers to obtain supplies of interferon at any price, the British government and the major funding bodies for cancer research in the UK issued a joint press notice in May 1980, cautioning against over-optimism and attempts to bypass the formal drug evaluation route.[38]

Despite the fact that the motives and enthusiasm of its own members had played an obvious part in the interferon mania, the scientific community swiftly left the media to carry the blame. For as long as the public enthusiasm interferon elicited was not associated with the term 'uncontrollable', the dramatic portrayal of interferon was widely accepted as necessary for the advancement of such a promising area of biomedical research. But once defined as a threat to the practice of medicine and professional autonomy, physicians on both sides of the Atlantic started making public appeals for a moratorium on publicity about interferon. The interferon community openly supported the physicians in their efforts to blame journalists for raising patients' hopes through irresponsible reporting.[39] The primary concern of the interferonologists was that the continued media frenzy, which had initially worked to the professional advantage of interferon research, might in the end rebound to discredit interferon research itself.[40]

It can be no accident that while doctors and authorities worldwide were struggling to cope with what they regarded as mass hysteria about interferon, preliminary results of the American Cancer Society sponsored interferon clinical trials, announced at the annual meeting of the American Association for Cancer Research, were unexciting.[41] The information made available cooled expectations by suggesting that interferon was no more active than other available chemotherapeutic agents in treating breast cancer and multiple myeloma. Moreover, contrary to what was hoped for (in particular by outsiders who were sold on interferon as a non-toxic agent), side-effects similar to those of other cancer medications were reported. The patient who responded best was said to suffer the most serious side-effects such as abnormal liver function and even cardiac toxicity, and therapy had to be interrupted.[42] These results placed interferon in perspective, just like any of the other many substances being tested for anti-tumour activity.

The redefinition of interferon as a potentially harmful experimental treatment with uncertain therapeutic benefits proved effective in undermining its public image as a wonder drug and put a lid on 'the unquenchable desire for interferon by cancer victims'.[43] Almost overnight the media picked up the message with headlines such as 'Interferon: studies put cancer use in doubt', 'Is it a wonder cure for cancer or the most expensive flop in history?'[44] Just as the interferon alliance had been overly successful in seeking a favourable press, the biomedical establishment managed to reshape the nature of media coverage of interferon – from forceful promotion of interferon's benefits to disappointment about its failed promise.

Despite the fact that interferon earned itself the dubious status of a 'miracle drug looking for a disease', it did not drop out of the picture to end its days on the laboratory shelf. A number of the drug companies, new genetic boutiques and research institutions with vested interests in bringing interferon to the medical market had apparently passed the economic point of no return. Confronted with the rapid erosion of public support, these interferon promoters helped to develop alternative strategies to legitimate work on interferon. It was the perceived unique capacity of interferon as a biological drug acting through the immune system, rather than its presumed non-toxic nature as a specific antiviral or anti-tumour agent, that was used to promote work on interferon. Part of this face-lift was to present interferon as having an important advantage over conventional therapies: it linked the clinics to advanced laboratory research in tumour biology, molecular biology and immunology. The rhetorical strategy that would prove effective in establishing a more permanent base for support in the 1990s can be dubbed as 'beyond interferon': picturing interferon as a first step advancing towards a medicine of tomorrow.[45]

## 'Doctoring' the media as part of modern medicine's expectations trap

This chapter shows that there are no simple, one-dimensional explanations for how interferon garnered headlines and achieved popular notoriety successively in the 1960s and the 1970s. Neither technical arguments about interferon's innovative therapeutic qualities (real or apparent) nor the interest of a specific social group suffice to explain the successive cycles of public enthusiasm and disappointment. In both cases a convergence of different historical factors made interferon temporarily into an acclaimed medical breakthrough.

In analysing Isaacs' early success at connecting his private science with public issues of disease and cure, in the British context, I located a strong public engagement with spectacular medical advances. In the post-Second World War period, antibiotics, steroids and neuroleptics had provided physicians with treatments never before seen. The wide and intense acclaim for this seemingly endless flow of new wonders at the disposal of doctors helped to establish in mass culture a firm belief in major therapeutic achievements as a hallmark of scientific medicine.[46] However, this condition was far from culturally specific and cannot explain, therefore, the specific British enthusiasm for interferon as the new 'antiviral penicillin'.

In his most stimulating essay 'Penicillin and the New Elizabethans' Robert Bud points out the powerful and multiform symbolic role of penicillin in postwar British culture not only as an icon of medical progress, but also of Britain's erratic research potential and of America's ominous industrial dominance. In line with Bud, I argue that making public parallels with penicillin helped to link interferon rhetorically with major issues in the popular consciousness in postwar Britain: pride in scientific and technological abilities, resentment at the impotence to capitalize on the innovative potential and jealousy of American success.[47] As such the very formation and use of the penicillin-associated imagery in the British media served to give high visibility to a seemingly promising medical discovery and boost public expectations about interferon's future as yet another wonder of modern medicine. In both tapping and stimulating the public sentiment of national interest, the propagation of the penicillin-oriented imagery played a pivotal role in both the first and second wave of British enthusiasm in the 1960s and 1970s.

The story of how interferon achieved popular notoriety as an anti-tumour agent in America in the 1970s reflects the preoccupations of a different society in a different epoch. The theme of the 1950s and 1960s had been the firm and unconditional belief in a scientific medicine that was synonymous with therapeutic success and progress. In the 1970s, however, it was precisely this mythologizing image of progress and cure, that had become inextricably bound up with modern biomedicine, that came under challenge.[48] The fast-growing American medical counter-culture got a political hearing for its critique of the technological-interventionist and reductive orientation of a high technology medicine that did not seem to have lived up to its promise. Cancer medicine, in particular, was challenged for being far less clinically effective than was officially asserted and even for being counterproductive in doing more harm than good. This challenge went hand in hand with attacks on the medical establishment for wasting taxpayers' money on developing new, questionable forms of cancer therapy with a scientific basis while neglecting the apparently promising search for the natural and organic.

As I have shown, in achieving a high degree of compatibility between interferon as an innovative approach to cancer therapy and this specific 1970s culture of American medicine, the interferon promoters succeeded in building support for their case. To the most influential interferon champion, Mathilde Krim, interferon represented as much marketing as scientific testing. She was well aware of the ever-growing importance of the mass media in generating public support and was very skilful at 'doctoring' the media. Krim played a pivotal role in shaping media coverage of research with the goal of promoting interferon. The interferon lobby lent journalists a hand to frame interferon both as a natural remedy and as the latest product of the most up-to-date scientific research long ignored by a conservative cancer establishment. Thus they succeeded in addressing both the American public's habitual fascination with medical novelties and its more fashionable appetite for unorthodox therapies.[49]

However, the role of the press went beyond journalists acting as mediators between the interferon lobby and the public: it often dramatised the accounts of interferon research in such a way as to captivate the attention of its readers. Notwithstanding the fact that media coverage of interferon reflected in part the expectations, legitimations and opinions circulating within the biomedical and public realm, the very mode of selection, emphasis and presentation of the information in the media played a significant part in shaping the cycle of promise and disappointment. Judging by the particular examples in this chapter interferon research offered ample opportunity to produce meaningful and arresting imagery which appealed not only to popular notions about health and medicine but also easily invited the audience to hyperbole: such as the magic and mystique of a scarce and costly natural substance with healing potential, the fearful association with a life-threatening disease, the promise of a medical breakthrough, the neglect of the unorthodox in scientific medicine, scientific and technological competition and the familiar parallels with the icon of medical progress and cure, penicillin. The media involved in the interferon story was of course a diverse and colourful enterprise which ranged from large circulation weekly news magazines, more specialized popular science magazines, to national and local newspapers, tabloids and, last but not least, commercial and non-commercial TV and radio networks. Yet a striking feature of the interferon reports is the homogeneity in their choice of imagery (metaphors as well as pictures) and terms. Apart from the fact that there was a tendency to mimic each other the imitative nature of the reporting can be related to a shared frame of reference which Anne Karpf has labelled the 'medical approach'.[50]

This implied the use of familiar medical icons such as the white-coated male doctor, the hopeful patient who is desperate for a cure, test-tubes and syringes, the high-technology laboratory and penicillin as the perfect example of the magic bullet. This kind of imagery gave to the interferon stories an aura of medical progress and the self-evidence of finding a cure, and however predictable in nature it made them understandable and appealing to readers and viewers. In addition, this medical approach meant that most interferon news reports were actually 'olds' (that is, not topical). Most interferon news reports preferred the human interest story and a few select glorious results or

dramatic events , such as single acute disease cases, interferon short-
ages, the heroic discovery, the race to clone the interferon gene and
seemingly miraculous cures. Wider social, political or economic issues
were neglected. Journalists pursued the apparent biomedical opinion-
leaders and rarely questioned the political or economic motives behind
their claims regarding interferon's therapeutic potential. However
predominant this rather uncritical medical angle it did not deter the
same reporters from covering the unorthodox part of the interferon
story with the same wonder and elan. With a keen sense of contemporary commonly held ideas about medicine they chose to adopt Krim's
projection of interferon both as a product of nature and of medical
research. The rhetorical use of the ambiguous term 'if' or 'IF' in 'mediating' interferon did everything but undermine the science-based
promise of a natural cancer cure.

But did the fact that interferon became the public embodiment of
scientific prowess and hopes of Elysian health suffice to turn interferon
into a miracle drug? Of course, as the promise of a cure for cancer
circulated between scientists, the media and the public and back again,
it gathered potency. However strong this self-enhancing inflationary
force it was not enough to amplify public expectations beyond
measure. It was the conjunction of a hoped-for cure with the promise
of a wonderful new technology, genetic engineering, which produced
an unprecedented global tidal wave of interferon-related publicity and
catapulted interferon into the elite league of miracle drugs. Neither
penicillin nor streptomycin nor cortisone could match the way interferon was rocketed to Elysian heights by a pounding cocktail of
modern mass media.

This brings us to the question of what characteristics these four therapeutic substances had in common that helped them to achieve
miracle drug status in the process of circulating between media and
public and back again. All four were experimental, new and initially in
short supply, and opened up the possibility of cure for life-threatening
or crippling diseases that had hitherto been incurable. In addition,
each had been involved in wonderful Lazarus-like healings at the
bedside. Furthermore, as the latest science-promoted promising products of modern medicine they all had 'natural' audience appeal as
breakthroughs and medical triumphs. Hailing them as wonder drugs
required a religious faith in a scientific medicine that was equated with
health and the quest for immortality. It is part of what Roy Porter calls
the 'rising expectations trap' of modern medicine.[51] But there is more
to the label of miracle drug than a religious hymning of medical

progress and cure. The miracle drug emblem is Janus-faced and transitory for a number of reasons: the fact that interferon was made of natural components facilitated its aura as an elixir of life, which was intimately linked to gullible beliefs in alternative medicine; there was also the association with the disappointing everyday experience of incurable disease and failing remedies.

Finally I will mention yet another repetitious quality to the interferon story. In the case of other miracle drugs like penicillin, cortisone and streptomycin a similar kind of deliberate creation of disillusionment occurred by a government-led coalition as a means to dampen public enthusiasm and stem the tide of public requests which it could not meet. An element of optimism in advertising a new remedy is widely regarded and even promoted as necessary for the advancement of medicine, as long as the public enthusiasm it elicits cannot be associated with the term 'uncontrollable'. My claim is that the limit of acceptable, controlled enthusiasm depended in each case on the extent to which the government and medical community feared a loss of authority over an increasingly empirical healing culture.

## Notes

1. D. Cantor, 'Cortisone and the Politics of Drama, 1949–1955' in J. Pickstone (ed.), *Medical Innovations in Historical Perpsective* (London: Macmillan, 1992), pp. 165–84.
2. A. Yoshioka, 'Streptomycin in Postwar Britain: a Cultural History of a Miracle Drug', in M. Gijswijt-Hofstra, G. M. van Heteren and E. M. Tansey, *Biographies of Remedies* (Amsterdam: Rodopi, 2002), pp. 203–28.
3. R. Bud, 'Penicillin and the new Elizabethans', *British Journal of the History of Science*, 31 (1998), 305–33.
4. 'Interferon may aid fight against flu', *The Daily Telegraph*, 16 May 1958.
5. Anonymous, 'Royal Society Conversazione', *British Medical Journal*, 1 (1958), 1229.
6. A. Isaacs to J. Lindenmann, letter dated 5 June 1958, Lindenmann Correspondence, Personal Archives.
7. For an extensive account of the early development of interferon, see T. Pieters, 'Interferon and its First Clinical Trial: Looking Behind the Scenes', *Medical History*, 37 (1993), 270–95.
8. R. A. Rettig, *Cancer Crusade* (Princeton: Princeton University Press, 1977), p. 18.
9. S. Panem, *The Interferon Crusade* (Washington, D.C.: Brookings Institution, 1984), p. 16; and interview with M. Krim, 11 November 1992, New York.
10. Interview with M. Krim, 11 November 1992, New York.
11. Panem, *The Interferon Crusade*.

12. J. T. Patterson, *The Dread Disease: Cancer and Modern American Culture* (Cambridge, MA: Harvard University Press, 1987), pp. 256–68.
13. Krim, interview.
14. 'Test Planned on Substance Used in Cancer Treatment', *The Washington Post*, 30 August 1978; 'Natural Body Substance: $2 Million Test on Cancer Retardant', *The Chicago Tribune*, 30 August 1978; J. Hixson, 'Interferon: the Cancer Drug We Have Ignored', *The New Yorker*, 4 September 1978, 59–64; 'New Cancer Weapon?', *Newsweek*, 18 September 1978, 90–1.
15. Krim, interview; interview with Frank Rauscher, 16 October 1992, Stamford, CT.
16. 'New Cancer Weapon?', *Newsweek*, 18 September 1978, 90–1.
17. K. White, 'Interferon: No Miracles Without More Molecules', *Medical Tribune*, 18 October 1978; A. Rosenfeld, 'If IF Works it Could ...', *Life Magazine*, July 1979; H. Lancaster, 'Potent Protein; Medical Researchers Say the Drug Interferon Holds Great Promise', *The Wall Street Journal*, 6 December 1979; F. Hauptfuhrer, 'Interferon: the Promising New Drug Against Cancer', *People* (US weekly), 2 July 1979.
18. M. Edelhart, *Interferon: the New Hope for Cancer* (Reading: Addison-Wesley, 1981), p. 146.
19. 'Interferon: the Body's Own Wonder Drug', *Saturday Review*, 13 October 1979; and 'Interfering with Cancer', *Scientific American*, April 1979.
20. The public announcement in early September 1978 of the cloning of the first medically significant human gene, to make human insulin, hit the headlines and was said to open up a new and most exciting era in biology. This only added to public interest in interferon: V. Cohn, 'Scientists in California Create Gene to Make Human Insulin', *The Washington Post*, 7 September 1978.
21. 'Race is on for Miracle Drug', *The Observer*, 30 December 1979.
22. Lancaster, 'Potent Protein', *The Wall Street Journal*, 6 December 1979.
23. This is nicely illustrated by a cover story of *Newsweek* magazine entitled 'DNA's New Miracles'; 'By turning bacteria into living factories scientists can cure disease and create new forms of life', *Newsweek*, 17 March 1980.
24. N. Wade, 'Cloning Gold Rush Turns Basic Biology into Big Business', *Science*, 208 (1980) 688–92; S. Andreopoulos, 'Sounding Board: Gene Cloning by Press Conference', *New England Journal of Medicine*, 302 (1980), 743–6.
25. 'Medical Breakthrough Reported', *Los Angeles Time*, 21 January 1980; 'L'interferon: Enjeu d'une Competition Mondiale Scientifique et Industrielle', *Le Monde*, 6 February 1980; 'Cancer Treatment Available Soon', *The Guardian*, 20 March 1980; 'The Big If–interferon', *The Listener*, 29 February 1980; 'Interferon: the IF Drug for Cancer', *Time*, 31 March 1980; 'Interferon Duurste Stof ter Wereld', *Telegraaf*, 29 May 1980; 'The Making of a Miracle Drug', *Newsweek*, 28 January 1980; *The Reader's Digest* monthly magazine with a worldwide circulation of about 50 million copies in fifteen different languages published the cover story 'Interferon, the new miracle drug?' in January 1980; K. Cantell, personal correspondence 1979–81; interview with K. Cantell, 4 May 1992, Helsinki; Interview with J. Sonnabend, 13 November 1993, New York.

26. 'Medical Breakthrough Reported', *Los Angeles Times*, 21 January 1980; 'Cancer Treatment Available Soon', *The Guardian*, 20 March 1980; 'Interferon: the IF Drug for Cancer', *Time*, 31 March 1980; 'At only $100 Million a Gram, This "Miracle" has a Future', *Science Digest*, April 1980.
27. Wade, 'Cloning Gold Rush'.
29. 'Wonder Drug Hope for Miss Anelli', *Sunday Mirror*, 1 June 1980; 'Dad's Wonder Drug Plea', *Sunday Mirror*, 8 June 1980; and 'Drug Brings Hope for Tumour Boy Daniel', *The Daily Telegraph*, 12 April 1980.
30. Thanks to Susanna Isaacs Elmhirst I was able to get hold of a copy of both the script and the videotape of the *TV Eye* documentary 'Cancer: the new weapon', 19 June 1980.
31. The use of the penicillin analogy was reiterative in the sense that journalists as well as interferon researchers made frequent use of it in their performances; K. Cantell, 'Why is Interferon not in Clinical Use Today?', in I. Gresser (ed.), *Interferon* (London: Academic Press, 1979), pp. 2–28, p. 3.
32. 'Publicity on Interferon Has Caused Great Distress, Specialists Say', *The Times*, 10 June 1980; J. Gordon McVie, 'Medicine and Media', *BMJ*, 28 June 1980, 161; R. Ridgway, 'Interferon: the Hopes and the Reality', *BMA News Review*, September 1980, 18–22.
33. Department of Education and Science to Sir John Eden (MP/House of Commons), letter dated 12 June 1980, MRC Archives File No. S806/5.
34. A flourishing black market in often dubious interferon samples developed rapidly, fuelled by those rich, famous and desperate enough to try anything, like the exiled, dying Shah of Iran.
35. MRC to ICRF, letter dated 23 June 1980, MRC Archives File No. D1009/40.
36. Internal note MRC, dated 23 June 1980, MRC Archives File No. D1009/40.
37. Interview with J. Petricciani, 6 November 1993, Cambridge, MA; and interview with D. Tyrrell, 21 May 1990, Salisbury.
38. Internal note MRC, dated 27 June 1980, MRC Archives File No. D1009/40; press notice, dated 13 May 1980, MRC Archives File No. D1009/40.
39. Editorial, 'What Not to Say about Interferon', *Nature*, 285 (1980), 603–4.
40. Edelhart, *Interferon: the New Hope*, pp. 1–9; 'Interferon: the Hopes and the Reality', *BMA News Review*, September 1980, 18–22.
41. 'Interferon Results "Promising", ACS will Commit Additional $3.4 Million', *The Cancer Letter*, 22 February 1980.
42. 'Interferon Results Cool Expectations', *The Cancer Letter*, 6 June 1980.
43. Edelhart, *Interferon: the New Hope*, p. 6.
44. H. M. Schmeck, 'Interferon: Studies Put Cancer Use in Doubt', *New York Times*, 27 May 1980; 'Is it a Wonder Cure for Cancer or the Most Expensive Flop in History?', *The Daily Star*, 19 June 1980.
45. T. Pieters, 'Marketing Medicines through Randomised Controlled Trials: the Case of Interferon', *British Medical Journal*, 317 (1998), 1231–3.
46. Cantor, 'Cortisone and the Politics of Drama', p. 173; H. M. Marks, 'Cortisone, 1949: a Year in the Political Life of a Drug', *Bulletin for the History of Medicine*, 66 (1992), 419–39.
47. Bud, 'Penicillin and the New Elizabethans', p. 305.

48. For a comprehensive account of medicine and the counter culture in Britain and America in the second half of the twentieth century, see M. Saks, 'Medicine and the Counter Culture', in R. Cooter and J. Pickstone (eds), *Medicine in the 20th Century* (Amsterdam: Harwood, 2000), pp. 113–23.

49. Bert Hansen argues convincingly that the high media visibility of therapeutic discoveries in America originates from as early as the 1880s: B. Hansen, 'New Images of a New Medicine: Visual Evidence for the Widespread Popularity of Therapeutic Discoveries in America after 1885', *Bulletin of the History of Medicine*, 73 (1999), 629–78.

50. Karpf's pioneering study examines the dynamics of mediating health and medicine by the mass media from the 1930s up to the 1980s: A. Karpf, *Doctoring the Media* (London: Routledge, 1988), pp. 9–31.

51. R. Porter, *The Greatest Benefit to Mankind: a Medical History of Humanity from Antiquity to the Present* (London: Harper Collins Publishers, 1997), p. 718.

# 13

## Afterword: Cultural Differences in Medicine

*Matthew Ramsey*

It is a sad privilege to add these brief reflections as a bookend to the late Roy Porter's characteristically generous and exuberant foreword, which will be among the very last of his many publications. The conference on which this volume is based was memorable in large measure for his participation – not only his paper, a survey of recent work on the history of the body, which has since appeared elsewhere,[1] but also his always illuminating and often provocative comments on the proceedings. Those contributions showed his customary versatility. Having come prepared to deplore the excesses of postmodernism and the misuses of theory, and finding a gathering of scholars similarly inclined (the ghost of Foucault barely stirred, and the theorist most often cited was Norbert Elias), he rapidly shifted tacks and pleaded for the considered use of theory.

The papers in this stimulating collection are as remarkable for their diversity and methodological eclecticism as for any common theme or argument. Yet taken together, they powerfully convey the diversity of human experiences of illness and healing and the multiple meanings assigned to the functions and dysfunctions of the human body. At least implicitly, many of the authors accept that meaning is located in particular social and cultural contexts. They stop well short of radical constructivism; there is no hint of formal standpoint epistemology,[2] or of Donna Haraway's theory of situated knowledges.[3] They still recognize, however, that there may be multiple ways of knowing, and that we need to be attentive to these variations, across time and space and, indeed, within the same society at any given time.

Do divergent meanings authorize us to speak of cultural differences and, indeed, of different cultures or subcultures in medicine? In his chapter on the eighteenth-century Catalan medical world, for example,

Alfons Zarzoso writes of 'a myriad of healing cultures'. A caveat is in order here. The anthropological literature has increasingly problematized the concept of culture, in two senses: the notion of a distinct meaning-producing realm of human activity, separate from the social or political; and a discrete set of symbols, rituals and discourses – 'a culture' – associated with a particular population group, set apart from others. The critics have emphasized the pervasiveness of meaning in social life, on the one hand, and, on the other, the extent of exchanges among human populations, even in relatively isolated parts of the non-Western world.[4] The contributors to this volume would no doubt agree. Their communities of knowers involved in sense-making are not fixed or rigidly bounded. Membership in one does not exclude membership in another; life experiences and the distinctive activities in which people engage shape meaning as much as collective discourses; and there is room for the idiosyncrasies of individuals, families and other small groups. One of the advantages of the term 'medical pluralism', which Zarzoso also employs and which most historians seem to prefer to 'medical multiculturalism' and cognate expressions, is that it is less suggestive of self-contained and mutually exclusive worldviews that might be identified with particular social groups or categories.

To emphasize differences in this way is not, of course, to deny the existence of widely diffused commonalities, such as the humoralist conception of human physiology, which has sometimes been invoked to support the idea of a single medical culture in the early modern West. Michael Stolberg's chapter on medical popularization and the patient in the eighteenth century recalls the common fund of knowledge shared by physicians and at least their educated middle-class and upper-class patients. Yet his finer-grained analysis of patients' experiences suggests that responses to disease were far from uniform: 'it was often from a mix of fairly heterogeneous, written and oral sources with frequently contradictory messages that patients and those surrounding them had to construct the most plausible interpretation of the history and symptoms of the disease and deduce the most promising practical approach'. Differences tend to stand out, even where there is considerable overlap, like regional idioms in a common language.

Each medical encounter between a patient and a medical practitioner or other care provider entailed both commonalities, without which the interaction could not have taken place, and differences, which might or might not matter greatly for diagnosis and treatment. When the divergence was substantial, conflict was possible. As Willem de Blécourt and Cornelie Usborne argue, negotiation to reconcile different meanings and reach some kind of consensus was a necessary prelude to treatment – though the consensus might be imperfect, leaving areas of disagreement or mutual misunderstanding. An analogous point could be made about a reader's encounter with a text. Here the live interlocutor is usually absent, though there were exceptions – one thinks, for example, of the many letters Tissot received from readers of his *Advice to the People* and *Onanism*.[5] Nevertheless, the text can be 'useful' only insofar as readers are able to appropriate its arguments and forge an understanding that accommodates their own basic views on health and illness.

De Blécourt and Usborne capture this dynamic of negotiation in the sub-title they chose for this volume, *Mediating Medicine*, and in their introductory chapter on 'Medicine, Mediation and Meaning'. They invoke two conventional meanings of the word 'mediation': transmission and reconciliation. The first might suggest a neutral medium conveying an idea from point A to point B. But even if the sender and the recipient speak the same dialect of the same language, messages can be altered in the process of transmission, in the fashion of a Chinese whisper ('telephone game' or 'gossip' in the United States). Different dialects require translation, and to translate is to some degree to transform. A key figure in several of this volume's chapters is the mediator, the go-between whose efforts at translation introduce the possibility of further reconfigurations of meaning. The second definition of mediation might suggest that compromise and a lasting settlement of differences are the norm. But mediation in the first sense can be an instrument of domination and delegitimation, or of resistance and counter-legitimation. With compromise or without, the relationship is in any case an unstable one. Consensuses evolve, or fall apart; disagreements, too, evolve along with the protagonists' changing beliefs and practices.

The mediators in the studies collected here bridge physical or social distance as well as differences in meaning. Mediators can be family members or neighbours, as in Yaarah Bar-On's chapter on early modern obstetrics and gynaecology. Hera Cook, in her study of British physicians who wrote sex manuals in the 1920s and 1930, sees them as using their own experience to mediate between the prevailing sexual culture and official medicine. Micheline Louis-Courvoisier and Séverine Pilloud, in their chapter on patients' correspondence with Tissot, focus on letters written by someone other than the patient. For Toine Pieters it is the media that mediated the meaning of interferon as a possible anti-tumour wonder drug. In Constance Malpas' elegant account of how the first orthopaedists exploited the aesthetic expectations and the corporeal self-consciousness of an emerging bourgeois elite, the mediators are surgical specialists seeking what we would now call self-referred patients in the medical marketplace. In a different vein, Catrien Santing examines the intersections of anatomy and Counter-Reformation spirituality in images of the heart, blood and circulation, with elite physicians providing a crucial link. Even Logie Barrow's analysis of the 'clashing knowledge-claims' of the supporters and opponents of smallpox vaccination in nineteenth-century England, the chapter that focuses most relentlessly on enduring conflict rather than rapprochement, is rife with examples of mediation, in which pro- and anti-vaccinationists pitch a certain interpretation of the procedure and its history to Parliament and the public.

The larger theme that emerges from the chapters, though, is that cultural differences in medicine can matter as much as the commonalities that make mediation possible. The various contributions reflect, as Porter notes, 'a new multicultural pluralism, sensitive to but not scripted by particular theoretical constructs'. This observation perhaps resonates with particular force coming from his pen. One (necessarily) selective reading of his massive oeuvre has emphasized his treatment of the common medical culture shared by physicians and an educated laity and more broadly by consumers in the rapidly expanding medical marketplace of Georgian England. Yet Porter, often polemical but never doctrinaire, did not make this context-specific interpretation an orthodoxy. The work of medical history for which he will no doubt be most widely remembered by the general public, *The Greatest Benefit to Mankind: a Medical History of Humanity*,[6] revisits the familiar arguments about medical commerce but also gives careful attention to the varieties of medical culture across the world and in the early modern and modern West. That catholicity of viewpoint and his keen awareness of the ways in which medical ideas and practices are situated in particular historical experiences were among Porter's great strengths. Recognizing them, together with his other better known and less easily emulated gifts and achievements, would be one of the most fitting tributes we could pay to his memory.

**Notes**

1. P. Burke (ed.), *New Perspectives on Historical Writing*, 2nd edn (Cambridge: Polity, 2001), pp. 233–60.

2. See, for example, N. C. M. Hartsock, *The Feminist Standpoint Revisited and Other Essays* (Boulder, Colo: Westview Press, 1998).

3. D. J. Haraway, 'Situated Knowledges: the Science Question in Feminism and the Privilege of Partial Perspective', *Feminist Studies*, 14 (1988), 575–99.

4. See W. H. Sewell, Jr, 'The Concepts of Culture', in V. E. Bonnell and L. Hunt (eds), *Beyond the Cultural Turn: New Directions in the Study of Society and Culture* (Berkeley, Los Angeles and London: University of California Press, 1999), pp. 35–61.

5. S.-A.-A.-D. Tissot, *Avis au peuple sur sa santé* (Lausanne: Grasset 1761); *L'onanisme: dissertation sur les maladies produites par la masturbation* (Lausanne: Grasset, 1760).

6. R. Porter, *The Greatest Benefit to Mankind: a Medical History of Humanity from Antiquity to the Present* (London: HarperCollins, 1997).

# Index